D1325958

SONGBIRDS

How to Attract Them and Identify Their Songs

SONGBIRDS

How to Attract Them and Identify Their Songs

CHRIS HARBARD

KINGFISHER BOOKS

A QUARTO BOOK

This edition first published in 1989
by Kingfisher Books, Grisewood & Dempsey Limited
Elsley House, 24-30 Great Titchfield Street, London W1P 7AD.

BRITISH LIBRARY CATALOGUING IN PUBLICATION DATA
Harbard Chris

Songbirds: how to attract them and identify
their songs.
1. Songbirds
I. Title II. Ord — Kerr, David
598.8

ISBN 0-86272-459-7

This book was designed and produced by
Quarto Publishing plc
6 Blundell Street
London N7 9BH

Senior Editor: Kate Kirby
Editors: Lydia Darbyshire, Steve Parker

Designer: Hazel Edington

Bird Artists: David Ord Kerr,
Robert Morton, Tim Hayward, David Hurrell
Illustrator: Vana Haggerty
Chart Composition: Elly King, Carol McCleeve

Art Director: Moira Clinch
Editorial Director: Carolyn King

Typeset in Great Britain by Text Filmsetters, London
and QV Typesetting Ltd
Manufactured in Hong Kong by Regent Publishing Services Limited
Printed in Hong Kong by Leefung-Asco Printers Limited

Contents

Introduction

BIRDS, MORE THAN ANY OTHER living creatures, have perfected the production of vocal sounds. The songs of birds are a great pleasure to man, inspiring both poetry and music linked with thoughts of nature.

Bird sounds can be heard on every continent, but not all their vocalizations can be called songs. Short and simple noises are commonly known as calls and can be produced by both male and female at any time of year. Songs are more complex and are usually produced by the male during the breeding season.

Of the 8,500 to 9,000 bird species in the world, one particular group is noted for its ability to sing – the oscines, or songbirds. These form 80 per cent of the order Passeriformes, which itself comprises more than 5,000 species of perching birds, or passerines, which range in size from small flycatchers to large crows, and which includes larks, nightingales, thrushes, warblers and finches. The remaining passerines are known as sub-oscines, and include the tyrant flycatchers, cotingas, antbirds and woodcreepers of the New World, Old World pittas and Australian lyrebirds.

Songbirds are aptly named for the sheer complexity of the songs they render, many of which are impossible to describe in words. They *can* be called warbling, twittering, trilling, rattling or whistling, or fluty, sweet, mellow, shrill, rich or liquid; but no combination of these words can truly evoke the sound.

The quality of song as we hear it differs widely among the songbirds, and what delights one person may not appeal to another. Certain species stand out, of course, and have provided inspiration to both music and literature. Of European birds, the nightingale

An eastern meadowlark (above), *a common bird of fields and meadows in North America, sings its sweet song from a fence post perch.*

'Fluent and wistful' are two adjectives used to describe the song of the willow warbler shown singing in a gorse bush (left).

A male blackbird (top) *shown in sleek breeding plumage.*

An African rufous-naped lark (above) *sings from the top of a thorn tree.*

inspires more myths and poetry than any other, perhaps the best-known poem being Keats' 'Ode to a Nightingale'. The songs of blackbirds, song thrushes, and skylarks have also been immortalized in verse. The repetition of a mockingbird's song is incorporated into Walt Whitman's 'Out of the Cradle Endlessly Rocking', and poems by the ornithologist Alexander Wilson describe many birds. In music, the nightingale makes frequent appearances, as does the cuckoo, in works by Beethoven, Haydn, Vivaldi, Liszt and Grieg. European goldfinches, linnets, swallows and robins can be heard in a number of musical works, and if you listen carefully to Dvořák's String Quartet in F, op. 96, sometimes known as the 'American Quartet', you can hear a scarlet tanager. The complexity of birds' songs makes their translation into music difficult, but one twentieth-century composer, Olivier Messiaen, has incorporated the songs or calls of 260 species from all over the world into 10 of his major works.

Almost concealed from view, a winter wren (right), *an uncommon North American bird, sings in an elderberry bush. For their small size wrens have loud and penetrating songs.*

Courtship of the common Asian iora (right) *includes an acrobatic song flight in which the bird makes a sharp ascent and sings as it spirals slowly down.*

A male reed bunting (left) *in boldly marked spring plumage at its song perch in a willow thicket.*

As the year progresses the bird song around us changes. Many of the first spring visitors are heard before they are seen, and the sudden addition of their songs to the sounds around us is always a source of great pleasure. Although song is merely a product of the reproductive urge, it seems to communicate much more. Spring bird song is part of the reawakening from winter, and the emotions it can generate in us make it easy to imagine that the birds are singing for sheer joy.

Reasons to sing

The reason some birds sing well and others do not is closely linked with their evolution and need to sing. Habitat plays an important part. In an enclosed environment – such as a forest where visual signals are difficult – loud song is essential for communication. A good example of this is the far-carrying tones of the South American bellbird. Forests contain a high proportion of highly vocal species and shelter some of the world's finest songbirds. Reedbeds and dense bushes are other closed habitats, and the warblers and wrens which live there all have loud songs. In a more open habitat, such as tundra or plains, birds can use visual displays more effectively; aerial songs here more commonly serve to draw attention to the display.

An analysis of those birds that primarily use song to communicate has shown that they tend to be small in size; have plumage which blends with their environment; use simple visual displays; live in a more enclosed habitat; have a marked territorial instinct; and are solitary breeders. Few birds show all these characteristics, but most songbirds show at least some of them.

INTERNATIONAL SONGSTERS

THE AFRICAN PLAINS are home to many birds sporting bright colours or elaborate plumage, both of which take the place of complex songs as a method of communication. Thus, widow birds have displays which make the most of their long tails and patches of bright plumage; many of the chats have coloured wing patches and display dances; and the starling species uses its iridescent colours to great effect. These birds usually have harsh songs, which serve solely to attract a female to the more dramatic visual display. Move to the tropical forests of Africa, however, and song becomes the predominant means of communication. One family of songsters includes the bulbuls and greenbuls, which are generally an uninspiring grey, green, brown and yellow. The robin-chats, another family of outstanding singers, live in dense forest undergrowth, and all have similar appearance, with orange-red underparts and a greyish back. The bou-bous, members of the shrike family, live in dense thickets and woodland and have evolved intricate song duets.

In Australia the same pattern is followed, with less musical birds in open country and the finest singers inhabiting the forests. Australia is home to some of the most brightly coloured finches – like the Gouldian finch, blue-faced finch and crimson finch – none of which has a song worth mentioning. In the forests there is a chorus of songs from pied butcherbirds, babblers, whistlers and honeyeaters, as well as from scrub wrens in the dense undergrowth. The best mimics in Australia must be the lyrebirds, brown, chicken-sized birds that live in the rain forest. To make up for their dull colours they have an extremely loud song in which they imitate the songs of other forest birds, combining them with their display, which makes use of their exceptional tail-feathers. The outer tail feathers of the male superb lyrebird are lyre-shaped and composed of alternate bands of light- and dark-coloured feathers.

When the male bird reaches the climax of its vocal renderings, the display is dramatically reinforced by the tail feathers being flipped over the head, the fine ones being vibrated to form a shimmering veil, which both entices and then engulfs the receptive female.

Rüppell's robin-chat (above), *one of East Africa's finest singers, has a warbling song and also mimics other species. It is a shy bird, found in dense upland forest, where it sings from thick cover.*

One of New Zealand's best forest singers is the bellbird (above), *which can be heard throughout the year. It has a number of different songs and at a distance some of the notes sound remarkably bell-like.*

Sometimes known as bellbird, the bell miner of Australia (above right) *lives in rain forest and gives a clear single note which varies in pitch. A group of them produces a pleasant tinkling sound.*

A pair of red-eyed bulbuls (right) *display aggressively to each other. Their loud and fluty song, typical of the family, can be heard coming from riverside bushes in southern Africa.*

In contrast, the birds of paradise, which mainly inhabit New Guinea, have what is possibly the most brightly coloured and ornate plumage of all birds. Their displays show off their plumes to the best advantage and may even involve hanging upside down. Their songs are disappointing, however, often just a soft warble.

Central and South America have more forested areas than perhaps anywhere else in the world. The tropical climate of much of it means there is little migration, as food is available all year round, and birds can successfully breed at almost any time of year. The dense rain forests can be divided into layers, with different species occupying each. Many birds on the forest floor, where little light penetrates and the ground is often open, are outstanding singers, such as the nightingale-thrushes. The numerous species of antbirds, although classed as sub-oscines, still have loud and, sometimes, musical songs. Most of the wrens – which live in forest undergrowth and thickets and are generally brown and skulking – are superb singers, with many of them duetting as a pair. Other members of the thrush family are among the world's best feathered singers, with the slate-coloured solitaire ranking at the top.

In the forest and thick scrub of Asia the most vociferous birds include bulbuls; babblers (especially the hwamei, a vocal and musical laughing-thrush); thrushes, like the shamas and robins; and flycatchers and warblers; many of these birds can be identified only by their songs.

A sedge warbler (right) *sings from the top of a flowering gorse bush. It produces a rapid medley of notes and, like its relative the marsh warbler, incorporates the songs of very many other birds.*

A red-throated pipit (above) *calls, camouflaged against grasses near its tundra nest site.*

The woodland species of Europe contain some excellent singers. The nightingale, in particular, illustrates the link between song and habitat well, being a basically nondescript brown bird, which habitually skulks in thick vegetation. Its song is not only complex but extremely loud and an inspiration to poets and musicians alike. All European thrushes are good singers, perhaps the least musical being the ring ouzel, which breeds in open upland hills and mountains and has a much shorter song phrase than its relatives. European warblers are mainly small brown birds, which live in woodland, scrub or reedbeds and have a delightful range of songs. Some, like the sedge warbler and the whitethroat, also have a song flight. The pipits also have song flights and complex songs, although nothing to rival that of the skylark.

No two bird species have exactly the same song, and wherever you travel there will always be new songs to hear. Various species earn the title of best singer in different countries. In Europe, it is arguably the nightingale or perhaps the skylark, although some prefer the songs of the closely related woodlark and

thrush nightingale; in North America, the hermit and wood thrushes and mockingbird take precedence; in South and Central America, the slate-coloured solitaire; in Asia, the shama and the Pekin robin; in Australia, the superb lyrebird and pied butcherbird; in New Zealand, the bellbird and tui; and in Africa, the robin-chats, especially the white-browed species.

Many birds are caged and bred for their singing ability, a practice that dates back to the Ancient Greeks. Finches of various types have proved the most popular as they are easily fed; and one species, the canary, is now bred in a truly bewildering variety of plumages. The large-scale trapping of birds for the cage-bird market still goes on around the world, and millions of birds are trapped each year; sadly, a high proportion of these die before they are even sold. Many countries have introduced tight controls on the export and import of birds, but many species – especially parrots – are seriously threatened by the trade.

The wood thrush (top) *is one of North America's best songsters, with an unhurried, liquid, bell-like melody.*

A mockingbird (above) *shown here at the edge of a Florida swamp, lives in a wide variety of habitats throughout North America.*

WHY BIRDS SING

A male red-winged blackbird (left) *defines its territory with its loud gurgling song, at the same time exposing its bright red epaulets.*

WHEN LISTENING TO A BIRD'S SONG it is often tempting to interpret its reasons for singing in solely human terms. Bird song gives us pleasure, but from the birds' point of view it is an entirely functional form of communication. Singing is directed at members of a bird's own species, and it follows that each species has a different song while individuals of each species sound alike. However, there is an occasional variation between different populations of the same species, in the form of dialects, which can distinguish sub-species.

In general it is the male that sings, although the females of some species can also sing, although usually not as well. Males use song to draw attention to themselves, whereas females in the vulnerable position of incubating eggs or young want to remain hidden.

Establishing home boundaries

One of the prime reasons for bird song is to establish the existence and boundaries of a territory. It is within this territory that a pair of birds will raise their young, so it must be jealously guarded from rivals as it will provide food and protection for both parents and young until the breeding season is complete. Singing lets other males of the same species know that a family is in residence and that intruders are not welcome. The song means nothing to different species, but as these will often have different food and nesting requirements anyway, a singer on its territory will ignore their presence, just as they in turn ignore his. Within a given territory it is possible to have robins, chaffinches, blackbirds, wrens and blackcaps all nesting without interfering with each other's needs.

The male chooses a series of song posts from which to sing, and by singing from each of them regularly he can define his territory. The borders are invisible, of course, but if ever a rival male crosses into an occupied territory he will be instantly challenged by the occupant. The ensuing fights or chases may sometimes involve aggressive bursts of song to help see off the intruder. The territory's original 'owner' invariably emerges as the winner of such disputes. Some cunning birds have an entire repertoire of songs and use a different one at each song post. This may well create the illusion of a number of occupied territories, and thus be more successful at keeping would-be intruders away.

Effective advertising

Another use for bird song is to advertise to a female that there is an unattached male present who is offering her a home. It follows, then, that a male who is paired up

will tend to sing less, using his song purely for territorial reasons, although how a female can distinguish between the songs of a bachelor bird and one that is happily paired is not clear. Singing also serves to strengthen the bond between a pair and may be involved with further courtship leading to mating.

In addition, some birds may have a sub-song, often a quieter version of their normal song. It does not carry for any distance and is usually heard only in the autumn and early spring. Some sub-songs are almost certainly rendered by young birds which have not yet reached the stage of full territorial singing, and they could therefore be regarded as a form of practising.

Bird calls are functionally different from songs and so have a different structure. They are short, simple sounds, usually of one or two syllables. Some are more complex, however, and almost fall into the song category, just as some songs are simple and – by the human ear, at least – could be mistaken for calls. Calls, however, communicate totally different messages from songs.

The robin (above) *sings throughout the year from both open and well-concealed posts. It is extremely aggressive and territorial.*

Starlings and house sparrows (right) *coexist in man-made built-up areas. Starlings tend to feed on insect larvae and worms and berries, while sparrows take mainly soft-bodied insects and seeds.*

A flock of male and female chaffinches (above) *feed on grain. In spring a loud call is used by young birds to find unoccupied territories and when they are selecting suitable song posts.*

A fieldfare (left) *displays aggression towards a starling. Loud calls often form part of such displays, which in this case relate to possession of an apple but could equally be over territory.*

A variety of calls

Songbirds tend to have a larger repertoire of calls than other birds, many having a vocabulary of 20 or more calls. Calls may be used to communicate with a partner, to beg for food, to call young birds, to keep contact with members of the same species within a flock, to show aggression or to signal that a predator is near.

Once the male's song has attracted a mate, calling between pairs often forms part of the courtship ritual. Courtship feeding by the male is usually accompanied by a begging call from the female, a sound similar to that used by young birds when trying to elicit food from their parents. Parent birds can call young birds to them when they have scattered after leaving the nest, which is particularly useful for precocial species (mobile before they are fully grown). Migrating birds call to one another when flying in a flock, a practice that enables a nocturnal migrant to rejoin the flock if it becomes separated. Feeding flocks in woodland call constantly; this helps them to forage more effectively and enables them to signal when they locate a food source. Winter flocks of birds like tits have foraging territories which may be defended from other flocks. This is particularly important in times of food shortage, when each member of the flock will identify itself by a call. If an individual who does not belong to the flock – and who therefore has a different call – enters the foraging area, it is immediately spotted and driven away. Threat calls form part of aggression displays when a territorial male is fighting off an intruder.

Perhaps the most interesting type of call is the one that alerts birds to the presence of a predator. The alarm call has the same basic pattern for a wide range of species. It is a short, high-pitched note which can be heard clearly, but gives little information about the location of the calling bird. This call is given when a bird of prey is spotted nearby and usually results in all of the small birds taking cover. A different type of alarm call will be sounded if a stationary predator is seen; owls usually provoke this type of call if they are discovered roosting. Instead of a short, high note, a loud scolding type of call, repeated many times, draws attention. This sound is interpreted by other species which usually come to investigate before joining in. The resulting chorus of scolding noises is part of a mobbing display, and the birds will often approach very close to the object of their attention. If the owl moves, the short, high call is given and all dive for cover. Mobbing usually persuades the target of the outburst to move on.

Its wings spread and tail fanned, a nightingale (left) *sings its legendary song from deep undergrowth. It gives a harsh scolding call when alarmed, and also a sweeter note as a contact call.*

How birds sing

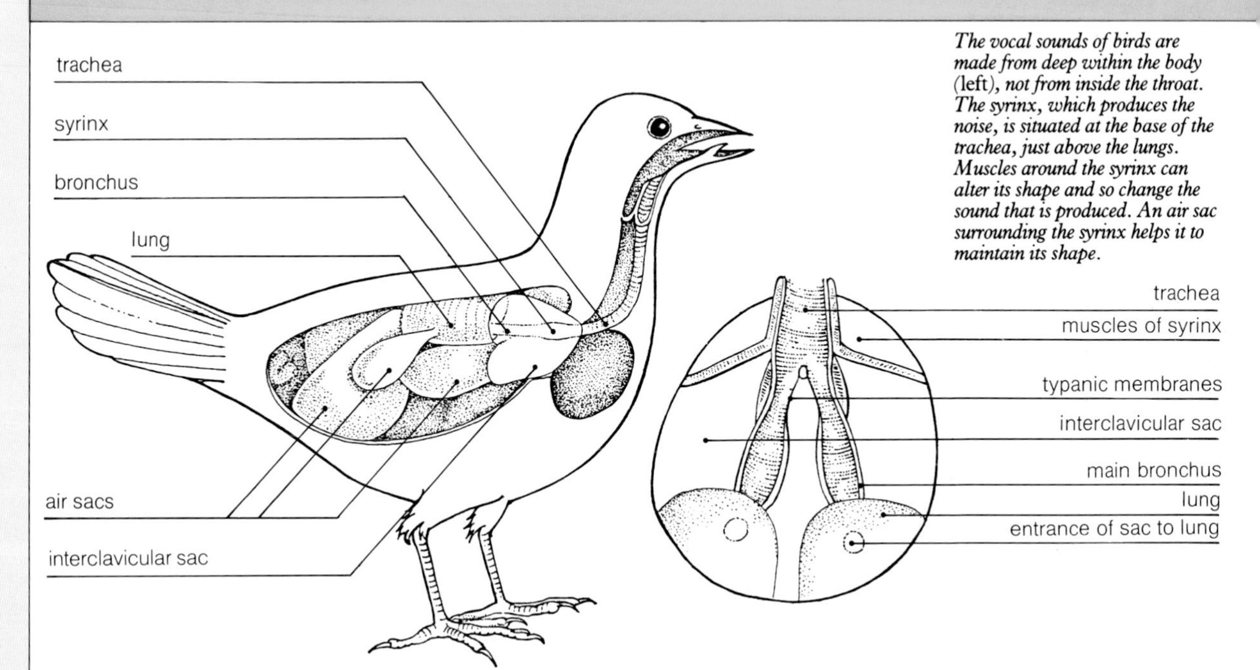

The vocal sounds of birds are made from deep within the body (left), not from inside the throat. The syrinx, which produces the noise, is situated at the base of the trachea, just above the lungs. Muscles around the syrinx can alter its shape and so change the sound that is produced. An air sac surrounding the syrinx helps it to maintain its shape.

THE VOCAL SOUNDS PRODUCED by birds are made in a completely different manner from those produced by mammals. Instead of having a larynx with vocal cords situated at the top of the trachea, birds possess an organ called the syrinx. This is a V-shaped structure situated at the base of the trachea (windpipe) where it divides into two bronchi that run to the lungs. Inside the syrinx are thin tympanic membranes which vibrate when air passes over them as it escapes from the lungs. The shape of the syrinx can be altered by muscles attached to it. This in turn changes the shape and tension of the tympanic membranes, which then vary the pitch of the sound produced. The more muscles to control the syrinx, the richer the range of sound. The principle can be illustrated by blowing up a balloon full of air and then stretching the neck of the balloon as the air escapes: the more the neck is stretched, the higher the sound. Vultures have none of these muscles, geese have only one pair, humming-birds have two, and parrots have three. Songbirds have from five to nine, which give them their wide vocal abilities.

VARIATIONS ON A THEME

OF THE SONGBIRDS, the ovenbirds, woodcreepers, and antbirds (the sub-oscines) have the most primitive type of syrinx, with the membranes attached only to the trachea. The rest of the songbirds have mem-

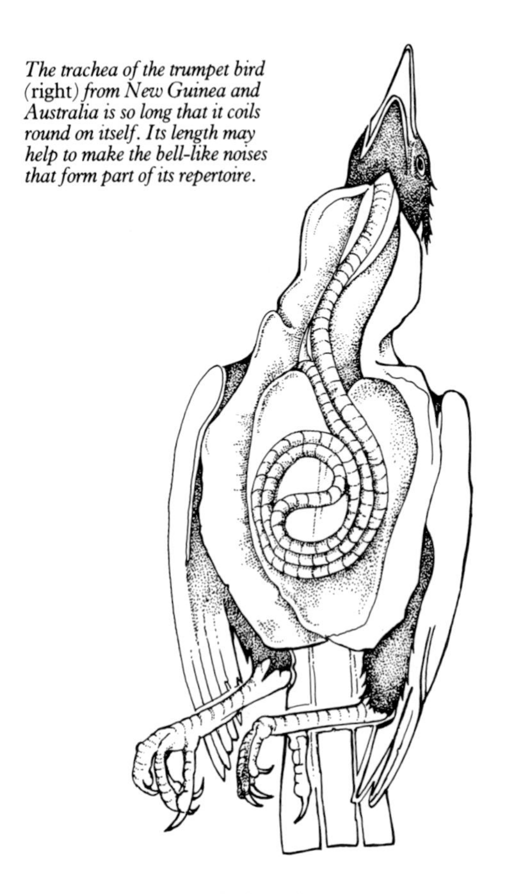

The trachea of the trumpet bird (right) from New Guinea and Australia is so long that it coils round on itself. Its length may help to make the bell-like noises that form part of its repertoire.

branes which can be attached to both the trachea and the bronchi in a variety of ways.

Part of a bird's respiratory system is formed by a series of air sacs; these allow a greater volume of air to be taken in than the lungs alone would hold. The interclavicular air sac surrounds the syrinx and exerts pressure on it. This particular air sac is essential for the production of sounds; if ruptured, no sounds can be made. The trachea can also play a role in sound-making by acting as a resonance chamber.

Among the different bird species, there is a great variation in syrinx structure and, therefore, the types of sound produced. Reed warblers can use the two branches of the syrinx independently, producing two notes at the same time.

The length and width of the trachea also play a part in the final sound. A short, narrow trachea produces a higher resonance than does a short, broad trachea. In the trumpet bird, a bird of paradise, the top and base of the trachea are only 3 or 3½ inches (8-9cm) apart, and the trachea, about 20 inches (50cm) long, coils around on itself. As song originates from the base of the trachea, it is possible for birds to sing with their bills full of food or even closed.

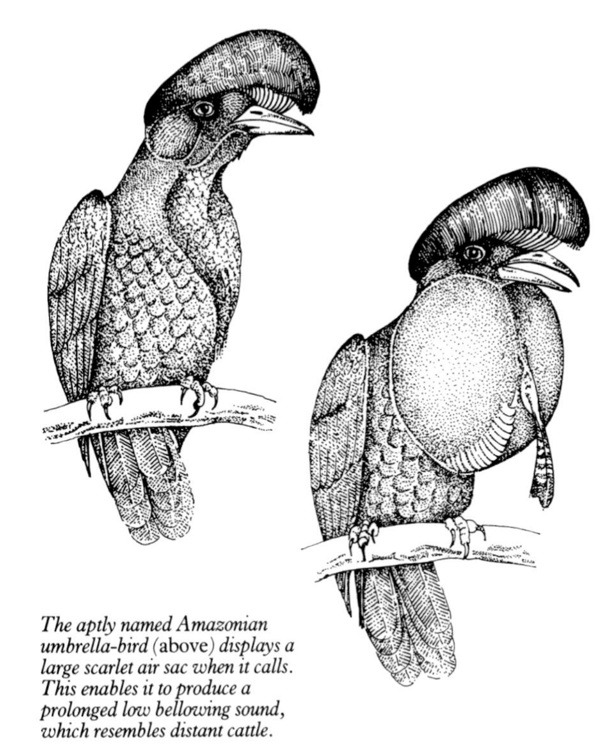

The aptly named Amazonian umbrella-bird (above) *displays a large scarlet air sac when it calls. This enables it to produce a prolonged low bellowing sound, which resembles distant cattle.*

Large 'sacs' or tympani, formed from an enlarged oesophagus, act as resonance chambers for the greater prairie chicken (left). *They are largest in the male, which inflates them during courtship, amplifying his booming call. It can be heard more than a mile away.*

Woodpeckers communicate over large distances by hammering with their beaks on dead trunks and boughs. The great spotted woodpecker (right) *produces a rapid, resonating drumming sound. The outer primary feathers of woodcocks* (far right) *have been converted into effective sound vibrators. During their courtship roding flights the vibrating feathers make a high whistling sound.*

Unusual sounds
Some species that are not noted for their songs use their air sacs to produce extraordinary sounds. The black grouse and capercaillies inflate theirs to produce a booming sound. Pigeons inflate their oesophagi with air to give their characteristic cooing tone. The modification of syrinx, trachea and air sacs is what makes bird song and sounds so beautifully varied and unique.

Many birds – although relatively few songbirds – produce sounds other than vocal noises by using parts of their bodies. The low thumping and drumming of the North American ruffed grouse is produced when it fans its wings to trap air next to its body. Hummingbirds are so-called because of the noise their wings make, and each species produces a different hum. Wing clapping occurs in the display of some pigeons. Common nighthawks produce a booming sound with their primaries. The North American woodcock has specially modified primaries which whistle. Of the songbirds, South American manakins make use of wing noises in their displays, producing a variety of loud snapping and whirring noises, while African flappet larks also generate wing sounds.

Other parts of the body can be used to produce sounds. Storks clap their bills together to make a loud rattling sound, and woodpeckers drum with their bills against a tree or branch. Snipe produce a drumming or winnowing sound with their outer tail feathers, and one songbird – the lyre-tailed honeyguide from Africa – is thought to make a sound with its outer tail feathers.

Learning songs
Just how much of a bird's singing ability is inherited and how much is learned was one question asked by early ornithologists. In the early eighteenth century, Baron Ferdinand von Pernau observed that if a young bird never hears the song of its parent it does not develop its full natural song. More recent experiments have shown this to be true, although the Baron was only partly correct when he also concluded that these young birds would learn the song of any other species that was put with them.

When a bird is isolated from the influence of its

Bill clattering of the white stork (left) *is the typical sound communication used between storks, especially when the sexes greet each other on the nest. The outer tail feathers of snipe* (far left) *are much stiffer than the central ones and are held together with tiny hooks. When the male snipe dives at an angle of 45° the tail feathers vibrate with a drumming or beating sound. The tails of pintail snipe have extra stiffened outer tail feathers.*

parents' sounds, it will produce a song that has the same basic pattern and sequence as an adult's song but that lacks the more difficult – and often more musical – parts. The full song is gradually built onto the basic pattern only after hearing the parents sing during the summer months. One experiment involving nightingales showed that in order to learn the complete song, a young bird had to see as well as hear the parent bird.

The amount of time taken to learn songs varies from species to species. Birds from temperate regions (where territoriality is strong) learn their songs at an early age, whereas in tropical regions (with less territoriality) the learning of song patterns goes on for a longer period of time.

Mimicry

There are some species in which the majority of the song appears to be innate. The North American song sparrow, for instance, does not need to hear any other song sparrow to develop a perfect song pattern. When it does hear one, however, it develops a song that mimics the bird it hears. As a result, song sparrows may have as many as 900 distinct local dialects.

As well as learning their own songs, many birds are excellent mimics, building the phrases of other species into their own songs. The Indian hill mynah is an accomplished mimic in captivity, to the extent that it can even copy human speech; in the wild, however, it does not incorporate the songs of other birds into its own. On the other hand, the marsh warbler, which breeds in Europe and winters in Africa, has a song that is made up almost entirely of phrases from other species. Studies have shown that these consist of a mixture of phrases from nearly 100 European species and over 100 African species. Each individual bird can mimic about 75 species; the most common European bird phrases come from the blackbird, house and tree sparrows, whitethroat and swallow, while the commonest African species are the common bulbul, green-backed camaroptera, black-backed puffback, tawny-flanked prinia and red-faced cisticola. The most notable North American mimic is the mockingbird, known to imitate 55 species in an hour. Catbirds and yellow-breasted chats are also good mimics.

When birds sing

IN TEMPERATE REGIONS the amount of daylight plays an important part in the life of all birds. It tells migrants when to migrate and it also triggers the production of hormones that prepare birds for the breeding season. Singing is stimulated by the presence of the male sex hormone testosterone, which is produced when daylight length reaches a certain amount – usually more than 12 hours. In temperate regions, this results in a distinct season during which the majority of birds breed. In the tropics, where the change in day length is minimal, breeding occurs throughout the year. As bird song is linked so closely to breeding, most singing takes place during the breeding season.

The length of song period varies from species to species and the timing of the breeding season differs geographically. Birds that breed in temperate regions have a clearly defined breeding season. Many birds begin singing in earnest at the beginning of the year in order to establish their territories early on. As spring approaches more and more birds sing, and with the arrival of spring migrants the chorus is complete. In the northern hemisphere there is nothing to compare with the woodland dawn chorus in early May.

For the most part, spring visitors will have been silent on their wintering grounds, but as they prepare to migrate they often begin to sing. The song is often very tentative to start with and far from perfect; many

Red-backed shrike

A male indigo bunting (above) *chooses a prominent perch to sing from. It is a summer visitor to North America and sings throughout the day from May until well into August, later than many other songbirds.*

Redstart

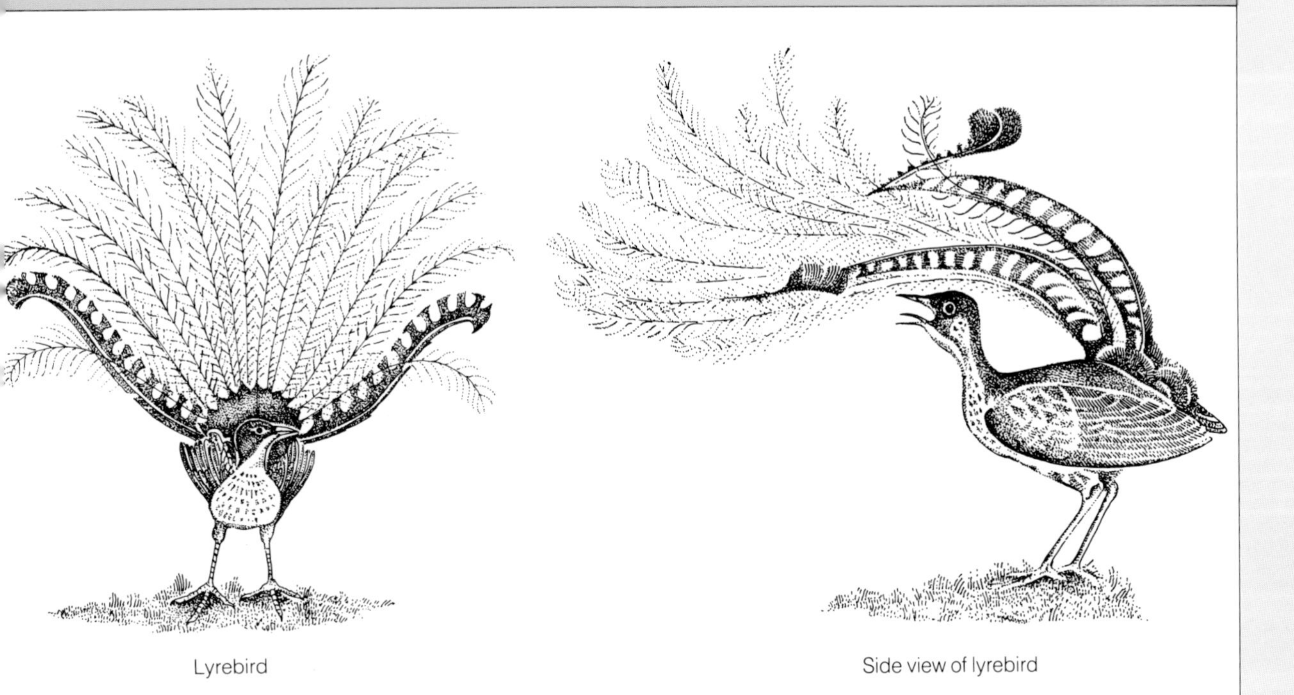

Lyrebird

Side view of lyrebird

Lesser superb bird of paradise

young birds will be singing for the first time. When these migrants eventually reach their breeding grounds they will be in full song, ready to establish a territory and find a mate.

PROMPTED BY DAYLIGHT

LENGTHENING DAYLIGHT HOURS encourage resident birds to sing; as the days get longer their singing becomes more pronounced and continues for longer periods. Wintering birds are often silent for the whole of their stay, but occasionally some of the late-leavers may be stimulated into song by a particularly spring-like day. Birds can be fooled by light levels, and some will begin to sing in the autumn when the daylight length matches that of spring. This singing soons stops as the days get shorter, however.

In most species, song is at its fullest at the start of breeding, with males singing throughout the day. It wanes slightly during courtship and mating, but picks up again during the incubation period, although males which help with the incubation will sing less. As soon as the eggs hatch and help is needed to feed the young, the male will sing less often, concentrating his territorial song on the area around the nest.

Many birds go silent and moult when their young fledge and need no more supervision. Others may have a second brood; if so, the male will sing again with renewed ardour, often re-establishing territorial boundaries. In the late summer or autumn the hormonal drive tails off, and many birds stop singing altogether, although this pattern of song is a generalization and some species do not conform. The mockingbird and cardinal, for instance, sing in every season as they defend their territories throughout the year, while the brown thrasher ceases singing immediately after mating.

As well as this annual variation in song, there is also a daily rhythm. Birds that are active during the day will begin singing before sunrise as the increasing light reaches a certain intensity. As the morning progresses, birds begin to quiet down; singing picks up in the late afternoon, continuing until nearly dusk.

Different species are triggered into song by varying amounts of light, and when listening to a dawn chorus separate species can be noted as they start. The order in which birds first sing is fairly constant, although it depends upon the birds present. In Europe blackbirds are usually the first to pipe up. Next may come the redstart or song thrush followed by the robin and wren. Turtle doves and willow warblers are the earliest singers of the summer visitors.

It is interesting to note that thrushes are commonly found to be the first singers in most countries. In North

The eastern phoebe (above) *is a North American bird common around gardens and farms.*

The red-eyed vireo (left) *has the distinction of being one of North America's most vociferous of songsters.*

A red-winged blackbird (above), *resident throughout the United States, sings silhouetted against the moon.*

The yellow-breasted chat (below) *is another North American bird that may sing its long bugling notes well into the night.*

America, the robin is often first, followed by the wood thrush. In India one of the first is the Indian robin; in southern Africa the kurrichane and olive thrushes are among the first. In New Zealand, with its high number of introduced European birds, it is once again the blackbird and song thrush which begin the chorus.

The contrast between the dawn chorus and the quiet of midday is truly astonishing when you realize that the birds are still wide awake, although once again there are exceptions, as certain birds sing consistently throughout the day. Blackbirds, wrens and nightingales are just a few examples. The North American red-eyed vireo holds the world record for the number of songs given in a day – a staggering total of 22,197!

Night-time vocalists

Night is not solely reserved for owls – it is often surprising just what else can be heard under the cover of darkness on a spring night. Mockingbirds are the North American equivalent of the legendary nightingale and can be heard regularly at night, together with yellow-breasted chats. Outbursts of 'ecstasy' singing from yellowthroats and ovenbirds add to the night chorus which, of course, also includes woodcocks and nightjars.

WHERE BIRDS SING

Birds of more open habitat often have a distinctive song flight. The tree pipit (left) *will fly from a perch high into the air and then deliver its song while floating down to the ground with wings outstretched in what is called its 'parachute' display flight.*

An exposed song perch allows songs to be heard more easily, reaching the ears of both rivals and potential mates. In an urban area with few natural song posts this blackbird (above) *has found a suitable alternative perch from which to broadcast its message.*

AS THE MAIN PURPOSE of song is to attract a mate and deter competitors, it follows that the further a song carries the better. For this reason many male singers will choose a prominent perch, such as the top of a tall tree, bush or rock, and sometimes an artificial song post at the top of a telephone pole or tall building. The song post may be positioned at the edge of a territory, the boundaries of which are often marked by a number of song posts. The actual nest is normally sited well within the territory and, especially with large territories, the male will sing from a perch well away from the nest to avoid attracting predators directly to it.

For bird watchers the fact that different singing birds choose different but consistent perching places helps to locate the bird. This is particularly true of many woodland species; nightingales sing from the undergrowth; willow warblers from middle heights and mistle thrushes from the tops of trees. The warblers probably show the greatest diversity of singing locations among European bird families – from reedbeds to scrub and woodland. When migrating they can sometimes be found singing in areas where they would not normally breed, and this can cause confusion. However, singing, feeding and nesting levels are not always the same; many thrushes that feed on the ground, for instance, will nest off the ground and sing from the top of a tree. Also, in open areas such as fields, prairie, tundra and desert, many birds deliver their songs during flight. Larks and pipits will 'perch' in the air to deliver their songs.

Birds that sing from exposed perches tend to have shorter song phrases, possibly to reduce the chance of a predator catching them unawares in mid-song. Certainly some of the longest songs come from species that stay well hidden, such as the grasshopper warbler, which has a trilling song lasting for more than two minutes. The possible extra vulnerability means that normally song perches are positioned where the singer can see a predator coming, and also near cover for escape. When birds are establishing song posts, they often try out a few, eventually settling for the safest one that lets them communicate to the widest audience.

The nightingale (above) *is a bird that breeds in dense woodland and thicket. To make its presence known it has a loud and beautiful song. To increase its chance of contacting a mate, the male habitually sings at night, when little else is making a sound.*

Weed stalks may provide a song post for a spring-plumaged bobolink (below)*, but he also has a fluttering song flight. Summer visitors to North America, the males arrive back several days before the females and fill the air with their tinkling song.*

IDENTIFYING SONGBIRDS

TO THE INEXPERIENCED bird watcher, the way in which a practised ear can pick out and identify bird sounds seems almost supernatural. Most beginners cannot imagine ever mastering the different sounds, but with practice it becomes relatively easy. Most people aware of birds will recognize the songs of many birds around them: starlings, robins, tits, skylarks and thrushes are some more familiar species and families. Learning to recognize the commoner species provides a basis from which to expand. Having learned the songs and calls of birds in your own garden, visit a woodland where many of these familiar birds will be present. Listen for the sounds you know and pick out any unfamiliar ones. Try to track down and identify the singer, remembering that where a bird sings from will often help with its identification. Also, try to compare any unknown songs with a familiar one, as this may often be a clue to the family of the bird you are hearing. Whenever possible, go out with a more experienced bird watcher who will be able to draw your attention to the similarities and differences of songs, and also pick out and emphasize the important phrases.

An additional help may be a recording of bird song, either on record or cassette. There are a wide variety to choose from – some will play all species in systematic order, others may select the commoner species, and still others present birds found in a particular habitat. All are a great help to the bird watcher who wants to become familiar with the bewildering number of sounds that birds make. The great advantage of recordings, of course, is that they can be played again and again, thereby providing you with the ideal opportunity to familiarize yourself with bird songs.

Translating bird song

Some bird songs are easier to remember than others since they lend themselves to verbal descriptions. Chiffchaffs say 'chiff-chaff' and the song of the yellow-

Bird watchers on the reed-fringed shore of Lake McIlwaine, Zimbabwe (left).

A bird watcher using citizen band radio updates a rare bird sighting. News spreads fast among the ornithological fraternity and rare sightings bring enthusiasts out from far afield to what could be a once-in-a-lifetime chance to glimpse an unusual bird (below).

hammer can be written as 'little bit of bread and no cheese'. Unfortunately, such descriptions cannot be found to fit all songs and calls. For example, one guide describes the American robin as having a 'loud, liquid song, a variable *cheerily cheer-up cheerio*', which gives a very good idea of the quality and rhythm of the sound but doesn't really help if you've never heard one before.

If you don't have a tape recorder, it is useful to familiarize yourself with a means of writing down songs and calls, as this may help to identify them later. Early attempts made use of musical notation, and while this is versatile enough to describe the notes, speed, pitch and duration of a song, few people have the ability to apply it accurately or to interpret it afterwards. Written descriptions are possibly the best way of all, although again, what you hear and write down may not correspond with what is written in a book. The most difficult parts to transcribe are often the consonant sounds, as it is usually the vowel sounds which contain the character of the song. And don't forget that as well as the sound, some idea of the length of each note or syllable is needed. Add to this an indication of whether successive notes are higher or lower, or whether a note is rising or falling, and quite a complicated system is required. One method uses a line above the written sounds to indicate length, rise or fall in pitch, and perhaps sound quality, such as a trill. Capital letters can be used to differentiate soft and loud notes.

Bird watching in the Florida Everglades (left).

Bird watchers using binoculars and telescopes on a beach maximize sightings of off-shore sea birds (above).

RECORDING BIRD SONG

MANY BIRD WATCHERS also become bird photographers, sometimes simply to record a particular sighting as a memento but often in an attempt to get better and better pictures, which might possibly be published. Similarly, many bird enthusiasts take up bird sound recording.

To get the best results, specialized and often expensive equipment is needed. With present-day technology, the large reel-to-reel tape recorders of the past have all but vanished, and nowadays portable cassette recorders and light, hand-held microphones can produce high-quality recordings. If you buy cheaply the results are likely to be poor, so it's always worth investing in quality equipment. The recorder's microphone is its most important feature, and it may be necessary to have a feature for eliminating background noise, especially in woodland, where other birds may be singing at the same time. There is such a wide choice of microphones and such a bewildering number of terms to contend with – do you want a condenser or dynamic type? omni- or uni-directional? with or without parabolic reflector? – that the best way to choose is to seek advice from a bird watcher who is already experienced in recording sounds.

The same applies to the type of recorder. Cassette recorders are the cheapest, and there are many to choose from. Reel-to-reel recorders are more expensive, but they allow you to edit tapes by cutting and splicing, and also to slow down recordings to analyse the sound. If you simply want to listen to what you have recorded, cassettes are most satisfactory. Remember that the frequency response is one of the most important features of a recorder, and if you want to record faithfully the high trills of warblers, the response should go up to more than 10,000 Hertz. Just as important, however, is that the equipment will stand up to the rigours of life in the field, so it must be both durable and waterproof. Be warned that bird sound recording can be addictive; many a bird watcher has all but forsaken his binoculars for a microphone and recorder.

The recording of bird sounds has had one major effect on ornithology – the invention of the sound spectrogram or sonogram, a method of recording bird

Out in the field an ornithologist is up early to record winter migrants (above). *The aluminium reflector for channelling the sound is painted green for camouflage. It is linked to a professional reel-to-reel tape recorder. Headphones are used to monitor the recording.*

Unobtrusive dress and movements are the keys to success in getting close to birds. After stalking a warbler in camouflaged clothing, a bird watcher focuses on an early spring migrant (right).

sounds visually as a graphical representation showing pitch and frequency. Sonograms, which tend to look like a series of smudges, are reproduced in some field guides. These can be interpreted with practice, although they are not of much use in the field.

ATTRACTING BIRDS

THE FRUSTRATIONS OF BIRD WATCHING in dense woodland – where birds can be heard but not seen – can often be great. However, birds' sounds and their reaction to certain sounds can be used to a bird watcher's advantage.

From the bird enthusiast's point of view one of the most useful types of bird behaviour is mobbing. By imitating the call of an owl, or better still using a tape recording, birds can be drawn by their instinctive mobbing reaction. Imitating alarm calls can also work in this way, as does 'pishing' and squeaking. These last two terms describe noises easily made by any bird watcher. The first is made simply by pursing your lips and blowing out, and, if repeated rapidly, the sound will usually attract warblers, flycatchers, thrushes and a whole host of other birds. A squeaking sound is made by kissing the back of your hand. Practice will soon show which noise gives the best response. Owl imitations are less easy and potentially harmful, as well, since the persistent use of an imitation or recording of an owl's call could cause a bird to abandon its territory. Even more likely to cause disruption in the breeding season is the use of a bird's own song to trick it into thinking there is a rival male in its territory. Birds respond so vigorously to this ploy that they have even been known to attack the offending tape recorder in their frenzy to drive off the intruder. Some nature

A pair of nightingales attack a stuffed cuckoo (above). *It was found that the cuckoo's head caused the strongest aggressive response in the nightingales.*

A party of ornithologists (below) *make their way through the Amazon Forest. In such closed habitats bird song features strongly as a means of communication between species.*

reserves now ban the use of tape recorders because of their effects on birds when used by a steady stream of bird watchers. Careful and limited use of these techniques, however, should aid the birder without disturbing the birds.

OUT IN THE FIELD

FAMILIARITY WITH SONGBIRDS is gained by many hours in the field hearing unfamiliar calls, tracking them down and studying the individual as it sings. As this is being done, many other fragments that will lead to birding competence are being collected. You will begin to understand habitat preference, activity hours for a specific species, preferred levels of activity within the forest for feeding and singing, the times of the year when the bird is most active and perhaps information on the bird's food preference.

The barest essentials are binoculars and a good field guide. On the market today there is an overwhelming selection of binoculars. Which sort you buy is often dictated by your pocket, but a few things should be considered. Prices may range from £25 to nearly £550! How can there be such a wide difference? As might be expected, quality is the main factor.

The very expensive models are optically perfect for viewing under a wide range of conditions, from very low light to looking directly toward a lighted subject. The way the glass is ground allows for maximum use of light under all conditions. They are usually waterproof and dustproof. Nothing can be more frustrating than having binoculars fog in a moist situation and prevent you from birding. The prisms of the expensive models are more securely anchored in place, and this prevents slippage in heat or when jarred and so prevents knocking the binoculars out of alignment. When you pay less many of these features are lost. In addition, the more expensive glasses are either designed or can be modified to focus extremely close, a feature that is a must for tropical or dense-thicket birding.

However, for £85 or so a good pair of binoculars can be had that will fit the average birder's needs. As for power, ask three people and you might get three answers. The standard, and a good choice, is 7 × 35 with centre and right-eye focus. The seven power will allow a very wide depth of field and you will not need to keep focusing and re-focusing. In open spaces, 10 power can certainly be useful when viewing over great distances, but in the 'up close' birding of some forests, 10 power can be very difficult to use. In addition, the higher the power the more likely that the vibration while you hold the binoculars will distort the view. Do not be fooled into 'the more power, the better I'll see the bird' syndrome. People with 15 and 20 power binoculars will find them worthless when using them out in the field. If you do a lot of low-light birding, 7 × 50 might be better to allow more light in under low-level conditions. Take the time and look through the binoculars you will purchase; they need to be right for you to make birding fully enjoyable.

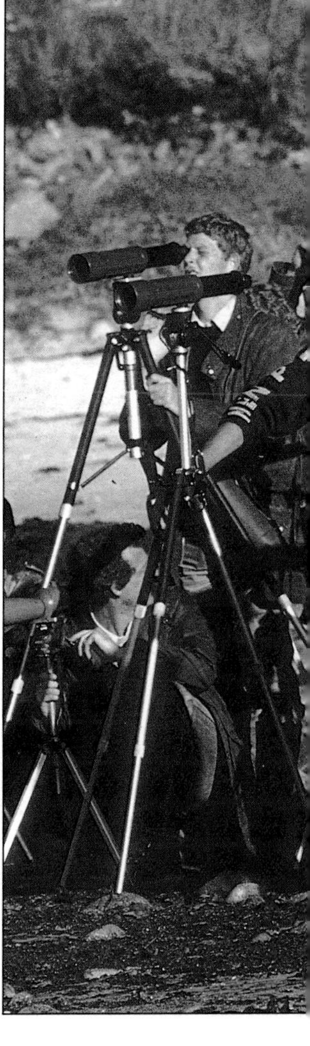

The excitement of the hunt gathers momentum (right) *as a dozen twitchers or more focus on a rare North American migrant blown far west out of its usual range onto the Isles of Scilly.*

Well prepared for a winter's day bird watching, a birder (left) *focuses his telescope on wading birds out on the mud flats.*

Telescopes

While we are on optics, the question of which telescope is best arises. Again, there are many makes on the market today. Most run in the £110-plus bracket. Here, too, it is important to look through the telescope before you buy it if you can. Some impart a greenish tint, some a bluish one, some are neutral. Different eyes view colours differently, and you need to get equipment that pleases your eye. Do not be fooled by the 'zoom to great power' feature. Once past 35 power it takes exceptional light conditions or one of the amazing £2000 level examples to make the high power worth it. In most telescopes the field of view becomes narrow and the image blurred. The ideal power is with an eyepiece of 20×, 25× or 30×. You can stay with one power and have as backup a 30× if needed, then just switch eyepieces. Generally, 20× can handle all needs.

When choosing a tripod be sure it is light and quick to set up. Some tripods have more levers and handles on them than the cockpit of a Boeing 747. Stay away from these! They stop you wanting to use the telescope. Birders may jump out of their cars, set up their telescopes and finish observing an area by the time a multi-handled tripod is just coming into place. It's back into the car and the complex job never did get set up. So go with light and simple equipment. If they're right, the 'scope and tripod will become a part of you in the field and you will carry them everywhere with no encumbrance.

GARDEN BIRD WATCHING

FOR THOSE PEOPLE WHO ENJOY the natural world around them, the presence of birds in their own garden can bring only pleasure. Enthusiasts lucky enough to have large gardens can entice a wide variety of species by providing the right environment, and even small gardens can be made attractive to birds. Many bird watchers get as much enjoyment out of a new visitor to their garden as they do from sighting the rarest of vagrants.

For a garden to be attractive to birds, it must provide sufficient food, breeding sites and protection. The most successful ones will reproduce something of the birds' natural habitat, especially those in which some parts have been left in a wild state to provide seeds and berries for food and thick cover for nesting and roosting. The garden should also be left untreated by pesticides so that a full community of insects and invertebrates can live there and provide further food. Such an area should have something to attract both breeding birds and winter visitors, ensuring a variety of birds throughout the year.

One of North America's larger garden birds is the blue jay (above). *Although omnivorous it eats mostly nuts and seeds. Oak and beech trees will provide it with food which it may carry off to store.*

The robin (left) *is one species that has adapted from a woodland to a garden habitat. Often tame and confiding, its bright colour, cheeky nature and sweet song make it a welcome visitor.*

PLANTING FOR SUCCESS

THE MOST ATTRACTIVELY MANAGED gardens for birds will contain: food plants, which can be a mixture of native and ornamental types; trees and shrubs, to provide further food, shelter, nesting sites, song posts and protection from predators; nest boxes, to provide suitable sites that do not occur naturally; a source of water, both for drinking and bathing (either a pond or a simple birdbath); and a winter feeding station, to provide a variety of foods (and which is visible from the house).

The food plants, trees and shrubs that are best for a bird garden will depend on the part of the country the garden is in, as well as its size. Native trees providing both cover and food are ideal. Oaks provide nesting places and cover and produce acorns. Beech trees produce nuts that are eaten by tits and finches; ash seeds are eaten by finches, and mountain ash berries attract waxwings and thrushes. Evergreens provide some of the best protection, while junipers also provide

A young cardinal (left) *perches on a spicebush. Berry-bearing bushes will help to attract birds into any garden. Some of the best are mountain ash, holly, elderberry, hawthorn and honeysuckle.*

Many flowers produce seeds for birds. The Eurasian goldfinch (below) *is one of the species that likes thistle seeds. Sunflowers produce seeds that are popular with tits, nuthatches and finches.*

berries, and the Scotspine forms cones which hold seeds for crossbills, tits and many other species. Berry-bearing bushes and shrubs are also essential – roses, hawthorns, holly, buckthorn and elderberry are some of the best. Of the shrubs, cotoneaster produces a prolific harvest of berries and comes in a wide variety of cultivated forms. Climbing plants such as ivy, honeysuckle and Virginia creeper are useful in smaller gardens, where they can grow up a wall to provide nesting and roosting places, as well as berries.

Birds that prefer low scrub and bushes can be enticed into a garden if a pile of twigs is made: a thick tangle of pruned branches and swept leaves should be left in a corner where birds can hide themselves. Also, in winter, the branches can harbour invertebrates for wrens and other insect-eaters to find.

Many attractive flowers produce seeds in winter for finches. Sunflower heads, for example, can be left on the plant, or collected and put out at a feeding station. Their seeds are great favourites of all birds, and are eaten by tits, nuthatches, greenfinches, sparrows and finches. Thistles, too, can be allowed to grow in a more secluded area to attract goldfinches, redpolls, siskins and other finches.

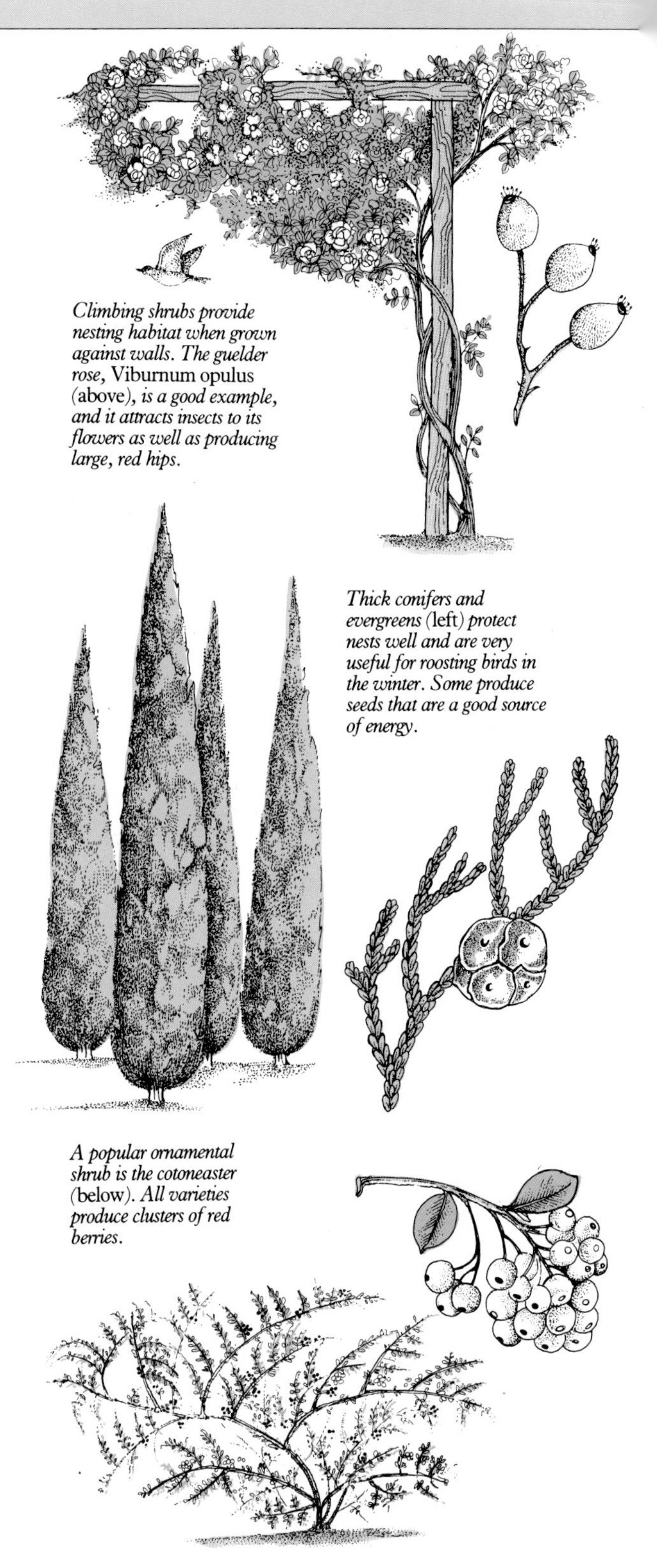

Climbing shrubs provide nesting habitat when grown against walls. The guelder rose, Viburnum opulus *(above), is a good example, and it attracts insects to its flowers as well as producing large, red hips.*

Thick conifers and evergreens (left) protect nests well and are very useful for roosting birds in the winter. Some produce seeds that are a good source of energy.

A popular ornamental shrub is the cotoneaster (below). All varieties produce clusters of red berries.

The oak (above) probably feeds and houses more birds than any other tree. As well as insects for warblers and acorns for jays, mature trees contain nest holes for nuthatches.

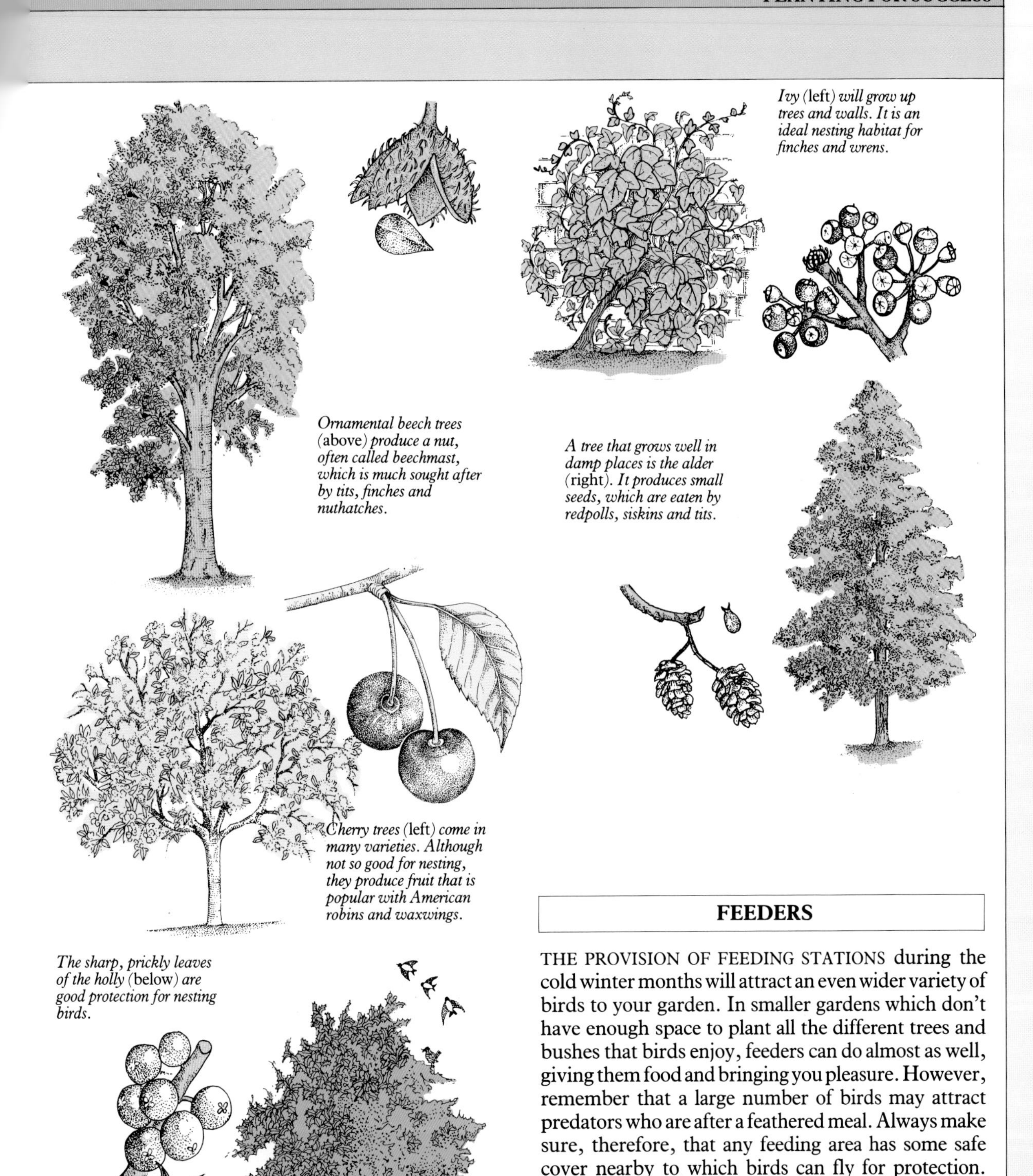

Ivy (left) *will grow up trees and walls. It is an ideal nesting habitat for finches and wrens.*

Ornamental beech trees (above) *produce a nut, often called beechmast, which is much sought after by tits, finches and nuthatches.*

A tree that grows well in damp places is the alder (right). *It produces small seeds, which are eaten by redpolls, siskins and tits.*

Cherry trees (left) *come in many varieties. Although not so good for nesting, they produce fruit that is popular with American robins and waxwings.*

The sharp, prickly leaves of the holly (below) *are good protection for nesting birds.*

FEEDERS

THE PROVISION OF FEEDING STATIONS during the cold winter months will attract an even wider variety of birds to your garden. In smaller gardens which don't have enough space to plant all the different trees and bushes that birds enjoy, feeders can do almost as well, giving them food and bringing you pleasure. However, remember that a large number of birds may attract predators who are after a feathered meal. Always make sure, therefore, that any feeding area has some safe cover nearby to which birds can fly for protection. Ideally, feeding stations should be situated out in the open where any birds can see a potential enemy approaching. If it is too close to a bush, a cat might hide there. But the garden itself away from the feeder should be planted with dense bushes and shrubs, especially evergreens, which provide protection from avian predators such as hawks.

Many birds forage over a large area for their food, and even birds that visit feeders may go to several gardens in turn. As a result, it is difficult to judge how many birds actually make use of the garden. You may see no more than a dozen blue tits at a time, but there could be more than 50 visiting each day. Their foraging instinct may be such that they tend not to feed in one place for long before moving on. Then again, some species, like starlings and chaffinches, will feed as long as there is food; they will rest near the feeders when the food is finished and feed again when more is put out.

A simple platform – a wooden tray fixed onto a post, window sill or tree stump – will bring many birds right up to your window. A roof is not essential but will serve to keep food dry. Feeders can be hung from the platform or from tree branches, and these can be used to dispense seeds, nuts and other food.

Ultimately, the key to a successful bird table is to put out as wide a variety of foods as possible. Proprietary seed mixtures contain different-sized seeds for different birds, for instance, millet, thistle, hemp, sunflower, corn and wheat. Peanuts and peanut butter are very popular – the latter can be smeared onto branches or put in log feeders. Fatty foods like suet provide birds with the most energy. They can be placed out in lumps, or melted and mixed with seeds to form a cake. Scraps of food such as cheese, meat fat and bones will also be gratefully received. Fruit is a favourite of many birds – apples, raisins, cherries, oranges and figs will all be

A seed hopper (above) *will gradually dispense seed into the tray and allow many days' supply to be put out at one time. It is especially useful if you are going away.*

A globe feeder (above) *stores a supply of seed and keeps it dry.*

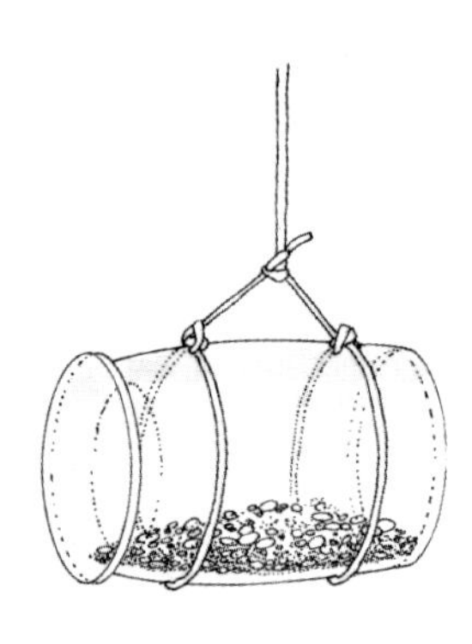

Instead of buying an expensive feeder, try using a suspended jar (above)*, which keeps the food dry and attracts tits.*

An empty coconut shell (above) *can be filled with a 'birdcake', made from seeds and melted fat, which has set before hanging.*

A tree stump (above) *can be used as a bird table by boring a hole in it and filling it with fat, seeds, fruit and nuts.*

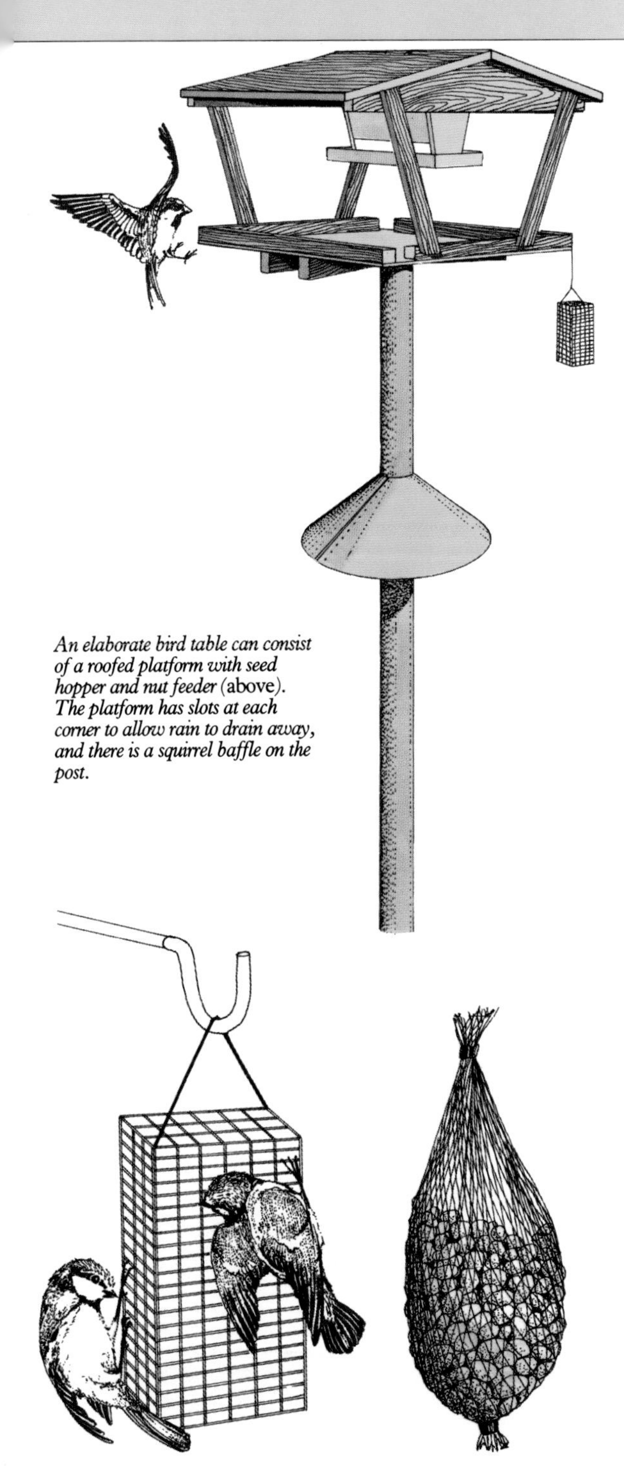

An elaborate bird table can consist of a roofed platform with seed hopper and nut feeder (above). *The platform has slots at each corner to allow rain to drain away, and there is a squirrel baffle on the post.*

eaten, and are excellent as summer food.

Bird feeders are used most after the breeding season, in the autumn, through winter and until early spring. Remember that while food attracts birds to where you can see them, it is only in severe winters that feeding may be vital. Also, feeding throughout the year is not always beneficial to birds. Many will leave the feeders as soon as natural foods become available again in spring, but some will continue to use them if food is present. Birds change their diets as the natural food sources change; those that feed on nothing but nuts and seeds in the winter will be eating insects and grubs in the summer. The young of these birds need soft insect food when they are developing, and they can be harmed if their parents give them unnatural foods from a feeder. Young birds may be unable to digest such food properly, and it might not contain all the necessary nutrition for their normal growth.

With large numbers of birds feeding, a considerable amount of droppings will accumulate around the feeding stations. These should be moved regularly to avoid contaminating the food of ground-feeding birds. Moreover, the accumulation of droppings can facilitate the passing on of diseases from one bird to another. If the bird table cannot easily be moved, it's best to disinfect the ground occasionally.

Cats and squirrels can be a great nuisance in a good bird garden. The former scare away the birds and occasionally catch them; the latter will steal food from bird tables and may also attack nestboxes. Feeding stations and nestboxes on posts can be protected by

A simple wire cage (above) *can be used for nuts and other scraps. Only agile birds can cling on to feed.*

A mesh bag filled with peanuts (above) *attracts finches and tits. Hanging it from a long string stops larger birds from feeding.*

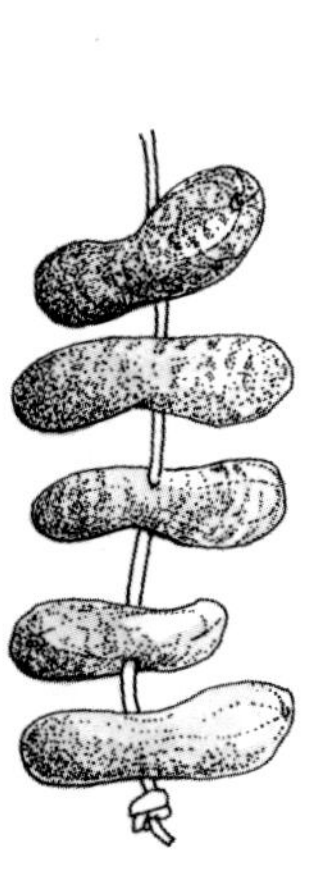

Peanuts in their shells (above) *can be threaded and hung out. It is fun to watch blue tits open the shells to get at the nuts.*

A small log (above) *can be drilled with holes and stuffed with suet and nuts. Woodpeckers will feed from it.*

using a baffle halfway up the post. Barbed wire is unsightly but often effective, although thorny cuttings from roses, holly or hawthorn, tied in a bundle, work just as well. A bell-collar on a cat may warn the birds of its presence and prevent their being caught. If a cat has a favourite position on a fence or under a bush, some prickly cuttings will soon persuade it to move on.

Some birds store the food they find. Nuthatches, some tits and jays will frequently take seeds, nuts and suet, which they then hide in a crevice or in the ground. They will return later for it – if it has not been stolen by other opportunist birds or squirrels that have seen them hide it!

It is debatable whether putting food out for birds during winter actually helps them to survive, except in the severest conditions. During a mild winter, the supply of natural food available will often keep birds away from feeders, especially if the trees have produced a good nut crop. Birds most need food when the temperatures are low and snow covers their normal food sources. Food is all-important to birds: it provides the energy they need to survive At night they lose energy in the form of heat, and during the day they lose more in the search for food. If they can't find enough food to replace the energy they have lost, they will eventually die. It is not just the provision of food that helps but also the fact that it is easily found and is there when needed. Food should always be placed out regularly during the winter months so that the birds can anticipate it and, if need be, rely upon it.

A good bird garden will also provide roosting places for birds so they don't need to travel far in order to feed when the weather is at its coldest. Sparrows and finches will simply perch deep within an evergreen, sheltered from the wind, rain or snow. In America a yellow-rumped warbler has even been known to roost each night on a Christmas tree in someone's house. Some birds will make use of a vacant nestbox; there is one case recorded of 50 wrens using the same box.

As well as simply looking at or listening to the birds in your garden, it can be interesting to study their behaviour, especially when feeding. Many exhibit a pecking order either within their own species or between different species. Among birds of the same species, males will often dominate females and not let them feed at the same time, while young birds may be driven off by older ones. Some species always appear more aggressive than others, displacing the more timid ones from the feeders. Occasionally, a bird may make a feeder, or even a berry bush, its own – mistlethrushes sometimes do this – and will drive off all intruders. If you notice that some birds are monopolizing a feeding station, try to spread out the food so that others can get to it. Don't just put seed in feeders or on a platform – throw some on the ground, as well, broadcasting it rather than placing it in one small area. This way all your garden visitors will get a chance to feed.

Providing water

During the winter months, water is as important to birds as food. They not only need to drink water, but must have it to bathe in. Birds' feathers are very important as insulation, and to perform efficiently they must be clean and dry. The air trapped underneath the

Beware – the versatile grey squirrel will raid even the most well-positioned bird feeder (right).

FEEDING AND NESTING REQUIREMENTS

Measurements of nestboxes are given as width x depth x height; nest hole dimensions refer to the diameter of the nest hole.

SPECIES	FOOD/FEEDING	NESTING/NESTBOX
SWALLOW	Insects caught in the air over water and fields.	Nests on beams, ledges and shelves in sheds and porches.
HOUSE MARTIN	Insects caught in the air over fields and water.	Builds mud nest under eaves; will use an artificial mud nest or build next to one.
WREN	Insects and spiders; visits gardens and takes small scraps and grated cheese.	Nests in sheds and sometimes in discarded objects; uses tit boxes for roosting in winter.
DUNNOCK	Shy feeder at bird tables; seeds and nuts.	Nests in shrubs and bushes.
ROBIN	Common visitor to bird tables for seeds, fruit and fat. Can become hand tame if fed mealworms.	Nests in any cavity in sheds and among vegetation; uses nestbox 10x10x20cm (4x4x8in) with top half of front open.
BLACKBIRD	Worms, insects and fruit; takes seeds, bread and apples from bird tables.	Nests in bushes, trees and wall creepers such as ivy and honeysuckle.
SONG THRUSH	Snails, worms and insects; takes fruit and berries from gardens.	Nests in trees and bushes, especially holly, hawthorn, conifers and creepers.
ICTERINE WARBLER	Eats insects and spiders; takes berries.	Nests in the fork of a bush or tree.
GOLDCREST	Small insects and spiders; visits gardens for aphids in autumn; may eat fat and bread.	Nests in tall conifers.
SPOTTED FLYCATCHER	Catches insects in flight.	Nests against tree trunk or wall creepers such as ivy; may use nestbox 10x10x20cm (4x4x8in) with upper half of front open.
PIED FLYCATCHER	Catches flying insects.	Nests in holes in mature trees; uses nestbox 10x10x20cm (4x4x8in) with 3cm (1⅓in) hole.
MARSH/WILLOW/COAL/BLUE TITS	Insects and seeds; visits bird tables for peanuts, sunflower seeds and fat.	Natural cavities in trees but blue tits will use almost any hole. Nestbox 10x10x20cm (4x4x8in) with 3cm (1⅓in) hole.
GREAT TIT	Insects and seeds; takes peanuts, sunflower seeds and fat from bird tables and feeders.	Uses natural tree holes and almost any other cavity; nestbox 10x10x20cm (4x4x8in) with 3cm (1⅓in) hole.
LONG-TAILED TIT	Insects, but occasionally visits bird tables.	Nests in thick bush or shrub.
NUTHATCH	Insects, spiders and nuts; very fond of sunflower seeds and peanuts at bird tables.	Nests in natural tree holes but will use nestbox 10x10x20cm (4x4x8in) with 3cm (1⅓in) hole.
TREECREEPER	Insects and spiders; may take crushed nuts from tree bark.	Nests in tree crevice and will use special wedge-shaped nestbox.
JAY	Insects, nuts and fruit; takes peanuts from bird tables.	Nests in mature trees and bushes in woodland.
MAGPIE	Insects, fruit and nuts; often visits bird tables.	Nests in trees and tall, thick hedges.
JACKDAW	Insects, snails and grain; visits gardens for scraps.	Nests in holes in trees,chimneys and nestbox 15x20x30cm (6x8x12in) with 8cm (3in) hole.
STARLING	Eats almost anything from a bird table or from the ground.	Nests in cavities in trees, roofs and walls. Takes over other birds' nestboxes.
CHAFFINCH/GREENFINCH/ GOLDFINCH/SISKIN	Seed eaters: chaffinches prefer mixed seed; greenfinches take sunflower seeds and peanuts; siskins eat only peanuts; goldfinches are attracted by seeding thistles.	Nest in trees and bushes.

Birds can be encouraged to breed in gardens by providing a suitable nest site. This spotted flycatcher (right) *is using a shelf which has been carefully placed against a wall behind a climbing rose.*

feathers provides insulation, which is why birds fluff out their feathers in cold weather. Regular bathing and preening enables them to clean and waterproof their feathers with natural oils. A supply of clean water should be available at all times and should be kept ice-free. Water is also equally important throughout the summer, when fresh water may be difficult to find. A simple birdbath will suffice for most birds, but a small pond with suitable vegetation near it will often attract birds that are otherwise difficult to spot in the wild. Never make a pond too deep, and be sure there are shallow edges. Branches laid into the water will provide perches.

A good bird garden should provide water as well as food. Regular baths (above) *keep this mistle thrush's plumage clean and waterproof.*

AN ASSORTMENT OF NESTBOXES

WHILE A GARDEN CAN PROVIDE relatively natural nest sites for birds which build in bushes, trees and grass, those birds which nest in tree holes rarely find a natural site. This is where bird boxes are indispensable. Boxes can be made to suit a wide range of species, but they should always be positioned in appropriate locations. When constructing a box, it is important to get the hole size correct for the species you want to attract. The box should have the right internal dimensions as well – not too big or too small. Lastly, the hole should not be too close to the floor nor too high up, although some birds may fill the box with nesting material until a suitable height is reached. Once built, the box must be positioned correctly, as some birds will nest higher than others. Most boxes are simply fixed securely to a tree trunk, although starlings, sparrows, tits and nuthatches will all use boxes, while swallows, spotted flycatchers and robins prefer open-fronted nestboxes. Remember that birds are territorial when breeding, and if you fill a garden with dozens of boxes it is likely that only a few will be used.

A robin (left) *carries nesting material to its nest, which may be in a nestbox, on the shelf of a shed or even in an old kettle. A good nest site is protected from both predators and weather.*

A nestbox should never be placed facing into the prevailing wind. If the box is tilted slightly downwards (below) *this will reduce the risk of rain getting in and chilling the nestlings.*

A nestbox (left) *is rarely too deep as the bird can fill the space with nesting material to its preferred height. Too shallow a box might allow a predator to reach in.*

Boxes should be cleaned out and repaired at the end of the nesting season. Removing an old nest will often remove parasites and their eggs that are lying dormant until the next breeding season, waiting to infest the young. If nestboxes are left up outside the breeding season some birds such as wrens, blue tits, long-tailed tits and nuthatches may roost in them.

Gardens that attract songbirds will always be full of activity throughout the year. Pleasure can be derived from watching the birds feed in winter, seeing them courting or squabbling over territories in the spring and raising their young in the summer. The arrival of the first migrants in spring and autumn can be most exciting and the occasional rare visitor will always be rewarding. A bird garden need not be untidy and overgrown and will contain more life than any carefully manicured, weed- and pest-free example of the horticulturalist's art.

A hole-fronted box (above) *is good for house finches, wrens and tits.*

An open-fronted box (above) *will appeal to some flycatchers. It should be partly hidden in foliage to keep away predators.*

Sometimes the urge to feed young can be too great. This blue tit (above) *has been attracted by the calls of baby wrens and is busy feeding them instead of its own brood.*

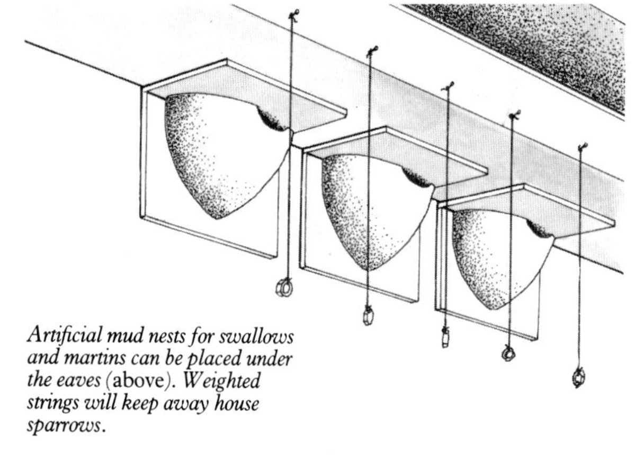

Artificial mud nests for swallows and martins can be placed under the eaves (above). *Weighted strings will keep away house sparrows.*

Lengths of dowelling are used for perches inside the roosting box so that many birds can enter (above).

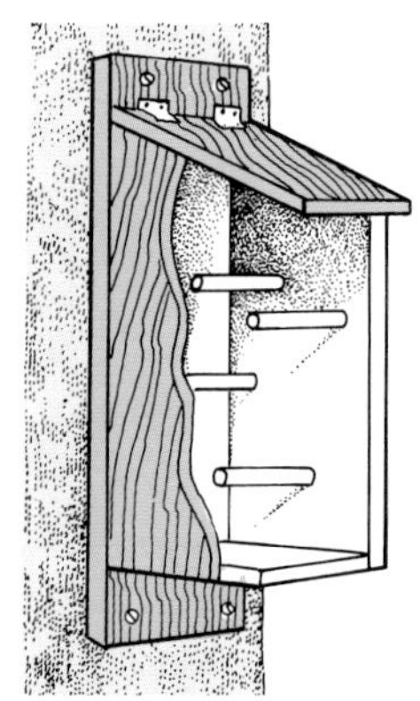

Some birds will use nestboxes as winter roosts. A roosting box can be made with an entrance hole and perch (above).

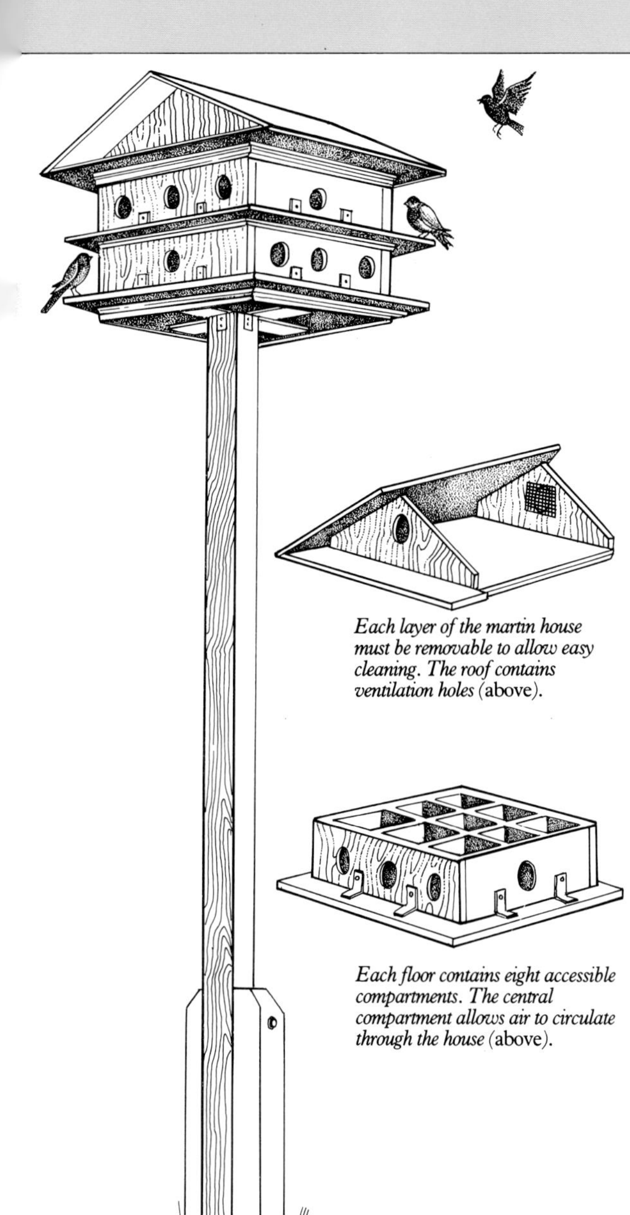

Each layer of the martin house must be removable to allow easy cleaning. The roof contains ventilation holes (above).

Each floor contains eight accessible compartments. The central compartment allows air to circulate through the house (above).

Summer visitors to North America, purple martins are colonial nesters and prefer an apartment block to a single nestbox (above). *Their high-rise accommodation is about 20ft (6m) from the ground and must be lowered for cleaning each year.*

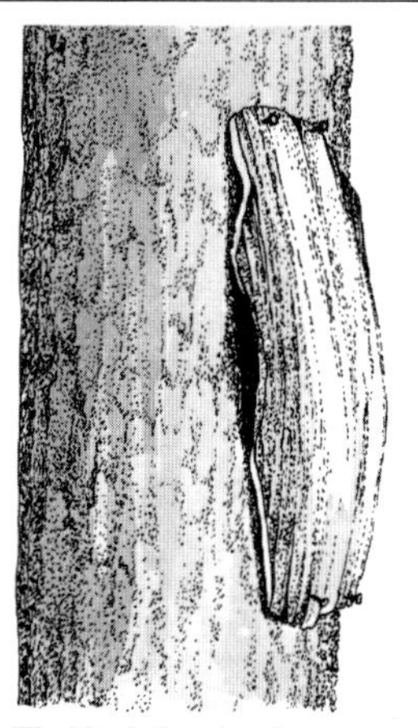

The North American brown creeper normally nests in a narrow tree crevice. A strip of bark fixed to a tree may be used for nesting or roosting (above).

A more open nesting tray with a roof (above), *may be used by American robins and phoebes if placed on the side of a house.*

All sorts of different nest sites may be used by wrens, such as a coconut with a hole in it (left) *or a hanging flowerpot* (above).

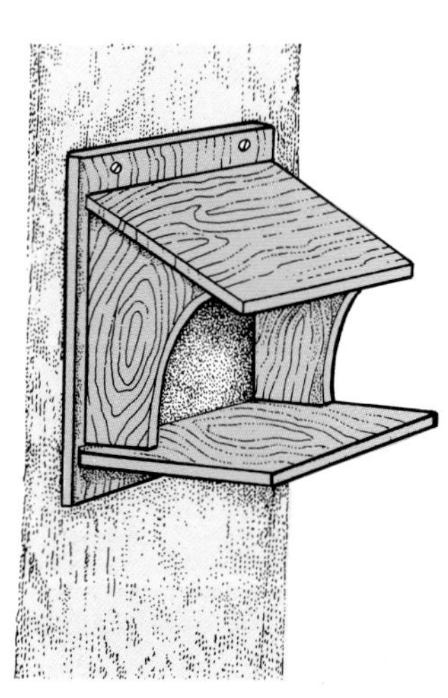

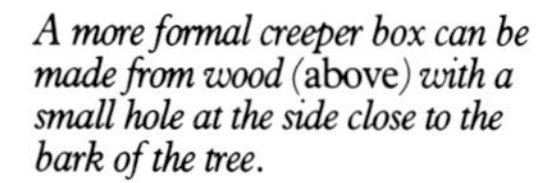

A more formal creeper box can be made from wood (above) *with a small hole at the side close to the bark of the tree.*

Birds need material to build and line their nests. A basket full of string, wool, hair and feathers (above) *will help them.*

THE DIRECTORY OF SONGBIRDS

How to use the directory

The great range of notes and rhythms on display in bird song can be confusing, but once mastered, the language adds a whole new dimension to bird study. This directory helps to do this, featuring the songs and calls of 100 songbirds, many of which can be attracted to your garden as breeding species. Winter visitors to the feeder, tempted by sensible natural food and cover, may stop by to feed or rest as well as those species on their spring and fall migrations.

Information on when and where all these birds sing is accompanied by interesting aspects of feeding and breeding behaviour.

With the birds grouped in families, it is easy to make out the broad outline of any common song features for each group – for example the sweet, clear notes of the thrushes, the buzzy, wheezy notes of sparrows, the rapid, scratchy songs of many of the warblers, and the song flights of the pipits.

- Summer distribution
- Winter distribution
- All year distribution

① Outlined maps of Europe give detailed distribution information, including migratory patterns. The maps show summer distribution (in green), winter distribution (in diagonal hatching) and all year distribution (in green with diagonal hatching).

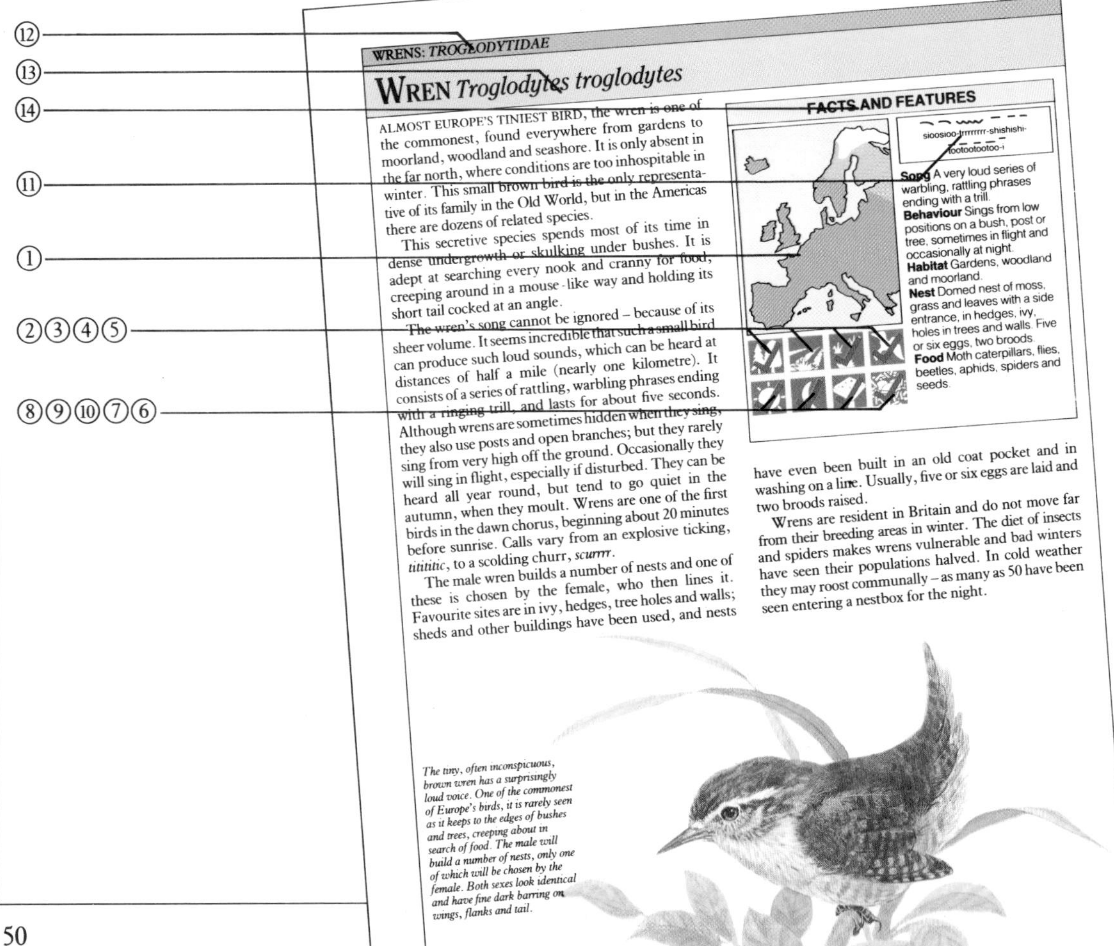

WRENS: TROGLODYTIDAE

WREN *Troglodytes troglodytes*

ALMOST EUROPE'S TINIEST BIRD, the wren is one of the commonest, found everywhere from gardens to moorland, woodland and seashore. It is only absent in the far north, where conditions are too inhospitable in winter. This small brown bird is the only representative of its family in the Old World, but in the Americas there are dozens of related species.

This secretive species spends most of its time in dense undergrowth or skulking under bushes. It is adept at searching every nook and cranny for food, creeping around in a mouse-like way and holding its short tail cocked at an angle.

The wren's song cannot be ignored – because of its sheer volume. It seems incredible that such a small bird can produce such loud sounds, which can be heard at distances of half a mile (nearly one kilometre). It consists of a series of rattling, warbling phrases ending with a ringing trill, and lasts for about five seconds. Although wrens are sometimes hidden when they sing, they also use posts and open branches; but they rarely sing from very high off the ground. Occasionally they will sing in flight, especially if disturbed. They can be heard all year round, but tend to go quiet in the autumn, when they moult. Wrens are one of the first birds in the dawn chorus, beginning about 20 minutes before sunrise. Calls vary from an explosive ticking, *tititic*, to a scolding churr, *scurrr*.

The male wren builds a number of nests and one of these is chosen by the female, who then lines it. Favourite sites are in ivy, hedges, tree holes and walls; sheds and other buildings have been used, and nests have even been built in an old coat pocket and in washing on a line. Usually, five or six eggs are laid and two broods raised.

Wrens are resident in Britain and do not move far from their breeding areas in winter. The diet of insects and spiders makes wrens vulnerable and bad winters have seen their populations halved. In cold weather they may roost communally – as many as 50 have been seen entering a nestbox for the night.

FACTS AND FEATURES

sioosioo-trrrrrrr-shishishi-tootootootoo-i

Song A very loud series of warbling, rattling phrases ending with a trill.
Behaviour Sings from low positions on a bush, post or tree, sometimes in flight and occasionally at night.
Habitat Gardens, woodland and moorland.
Nest Domed nest of moss, grass and leaves with a side entrance, in hedges, ivy, holes in trees and walls. Five or six eggs, two broods.
Food Moth caterpillars, flies, beetles, aphids, spiders and seeds.

The tiny, often inconspicuous, brown wren has a surprisingly loud voice. One of the commonest of Europe's birds, it is rarely seen as it keeps to the edges of bushes and trees, creeping about in search of food. The male will build a number of nests, only one of which will be chosen by the female. Both sexes look identical and have fine dark barring on wings, flanks and tail.

② The forest and woodland symbol indicates that the bird is usually found in treed areas, including orchards and gardens.

③ The open country symbol includes grasslands and bog.

④ The reed/cover symbol indicates that the bird is usually found in low vegetation.

⑤ The flight symbol indicates that the bird sings in flight or has a typical flight song or call.

⑥ The nest symbol indicates that the bird can be attracted to nest in gardens with suitable vegetation or a nestbox.

⑦ The bird table symbol indicates that the bird can be attracted to a feeding area.

⑧ Sings in day.

⑨ Sings at night.

⑩ More than one 'time of day' symbol may be ticked, for example when a bird may sing in the day, but predominantly at night.

⑪ Rhythm box; the rhythmic flow of a bird's song is given for each species using simple linear notation. The length of pause between notes depicts the rapidity of the song.

The base components are as follows:

Rising pitch

Drop in pitch

Long note

Fast rising inflection and sudden drop

A trilling effect

Very rapid

Long notes and long pauses

Short notes and long pauses

⑫ Family name, English and scientific.

⑬ Species name, English and scientific.

⑭ 'Facts and features' box for easy reference, which contains a distribution map, song description and pattern, behaviour, typical habitat, and nest and food information including types used to attract birds to the garden.

E

lla modularis

The Dunnock always appears nervous, walking or hopping with its curious shuffling gait, close to cover and ready to disappear if danger threatens. Dunnocks frequently chase one another during the spring, uttering insistent piping calls. Male and female look identical, with grey head and underparts, brown- and black-streaked back and pink legs. The song is similar to that of the wren but more tuneful.

ND FEATURES

weeso, sissi-weeso, sissi-weeso, sissi-weeso...

Song A high musical warbling.
Behaviour Sings from the top of a hedge, bush or tree.
Habitat Gardens, hedges and bushes.
Nest Of twigs and moss with leaves and roots, in hedges and evergreen bushes. Four or five eggs, two or three broods.
Food Beetles, caterpillars, spiders and worms. Seeds in winter.

ilso known as the hedge sparrow, is a bird in most of Europe. It is shy and close approach, appearing from a n with a grey head and belly. Closer is a beautifully patterned chestnut-reddish eyes and a slim dark bill, quite ue sparrow. In fact the dunnock is not sparrow family, it is an accentor. One ourful local names is 'shufflewing', accurately describing the bird's habits. It spends much time on the ground, walking or hopping with a distinct shuffling gait, flicking its wings as it goes. It is found in woodland, hedges and moorland as well as gardens and parks. This is a solitary bird except when feeding.

The dunnock's song is a high, cheerful warbling, similar to a wren's but more musical. Males usually choose a perch on top of a hedge or bush, or sometimes on a tree branch. Singing occurs throughout the year, but begins in earnest in late January or early February. The call notes are a clear, insistent *tseeep* and a high, rapid trill.

The nest is well hidden in the middle of a hedge or bush, often in a conifer or other evergreen and sometimes in thick scrub. The species is a favourite host of the cuckoo; unusually, the cuckoo is unable to mimic the plain blue eggs of the dunnock, yet the latter appears not to notice the difference. The sight of a parent dunnock feeding a cuckoo fledgling many times its size, perched on its back, can look very comical. Dunnocks have an interesting social life since they do not always form pairs when breeding. A single male or female may have many partners, or a number of males may share several females.

Dunnocks are resident in Britain, rarely straying far from their home territory. On the Continent the northern populations migrate south for winter and some of these are seen along Britain's east coast in the autumn. They eat a varied diet of seeds, insects, worms and other invertebrates.

69

CRESTED LARK *Galerida cristata*

The long, pointed crest of this lark distinguishes it from a skylark although its plumage is similar. In flight it lacks the white trailing wing and tail edges of the skylark, showing instead rounded wings with orange-buff undersides and buff outer tail feathers.

ONE OF MANY similar small, brown birds, the crested lark is found over much of continental Europe. A bird of open places, it is most commonly seen flying up from beside a road and fluttering lazily into a nearby field. It prefers drier areas than its relative, the skylark, and is less frequently found in mountainous terrain.

In common with many birds of open country, crested larks are well camouflaged, their brown plumage blending in well with the dry earth. If alarmed they may crouch close to the ground and become virtually invisible, only being noticed when they run or fly away. In flight, crested larks have broad, rounded wings and a short tail; in strong light, their plumage has a more sandy, even gingery look, compared with the skylark. Wings and tail lack the white edges which are a clear feature of skylarks.

Crested larks live up to their name, having a distinct conical crest which remains conspicuous at all times. They sing from the ground, often perched on rocks, and from buildings and bushes. Like so many of the lark family they also have a song flight. The song is usually delivered at between 100 and 200 feet (30 and 60 metres) above the ground and – unlike the skylark – there is no song as the bird flies to this height.

Crested larks begin to sing in January, but their main vocal period is from March to July, with further singing in the autumn. The song is not a continuous cascade of notes, like the skylark, although it does retain some of the skylark's qualities. The call is a three- or four-syllabled phrase, *whee-whee-wheeoo*, whistling and liquid, that varies in loudness and sweetness. The bird's alarm note is a low whistle, rising at the end.

The nest is built in a depression on the ground, usually by a bush or tussock of grass. Nests built in northern Europe tend to face north, away from the sun. Three or four eggs are usually laid and two broods are raised, with extra eggs farther south. The crested lark's stout bill testifies to its preference for a diet of grains and seeds, although in the breeding season it feeds its young on caterpillars and insects.

Crested larks are resident over their range, which extends eastwards to southern China and south to central Africa.

FACTS AND FEATURES

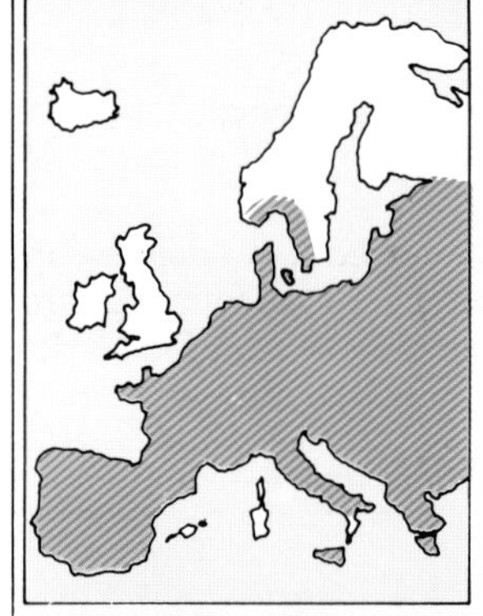

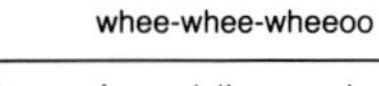

whee-whee-wheeoo

Song A warbling series of notes. Call notes include a whistling *whee-whee-wheeoo*.
Behaviour Sings in flight from high in the sky or from the ground. Walks and runs along the ground; has an undulating flight.
Habitat Barren and open country, arable fields, sandy and rocky ground.
Nest Of dead grasses on the ground in small depression. Three or four eggs, two broods.
Food Oats, wheat, barley and other grains and seeds.

WOOD LARK *Lullula arborea*

THE WOOD LARK was once widespread and common across Europe, but it is becoming scarcer and its range is contracting. The reasons behind its decline are not clear, but habitat loss and climatic change play their part. This bird likes lowland heaths, where it feeds among the short grass and perches on bushes or small trees, but it is also seen on rocky hillsides. In Britain it is often found where conifers are being replanted.

This lark is shorter than the skylark and is brown streaked with black – good camouflage for its surroundings, like most ground-nesting species. It has a distinct, pale-buff stripe above the eye, while the short tail is tipped white. The small crest is inconspicuous except when raised.

Wood larks sing for most of the year, but particularly from March to June. They are heard mainly during the day, but they also sing at night, especially when the moon is full. The song is more melodious and slower than a skylark's and contains repeated phrases, in particular a rich *lu-lu-lu*. They will sing from the ground or from a prominent tree perch. Males also have a typical song flight which differs from a skylark's in that the male wood lark flutters more as he rises from the ground, spiralling upwards and circling at a fairly constant height as he sings. When calling from the ground he gives a melodious *titlooeet*.

The wood lark nests on the ground, in a depression sheltered by grass or bracken. Three or four eggs are laid and two broods are normally raised. Wood larks feed mainly on insects such as beetles, flies and moths.

A short-tailed lark, the wood lark has a distinctive broad buff stripe above the eye and its brown plumage is streaked with black. Its crest is noticeable only when it is raised. On the ground the wood lark shows a black and white patch at the edge of the wing. In flight the tail has white tips but no pale outer feathers. The wood lark is to be found more often perched on trees and bushes than other larks.

In autumn and winter they also eat seeds.

In the west of its range this species is mainly resident, although it may move from its breeding area in early autumn. In the north of its European range there is a migration to parts of southern Europe in the winter, and some continental birds occasionally cross the North Sea to Britain. British breeding birds stay within the country except during severe winters, when some may move into France; they return to their breeding grounds in late February and early March.

FACTS AND FEATURES

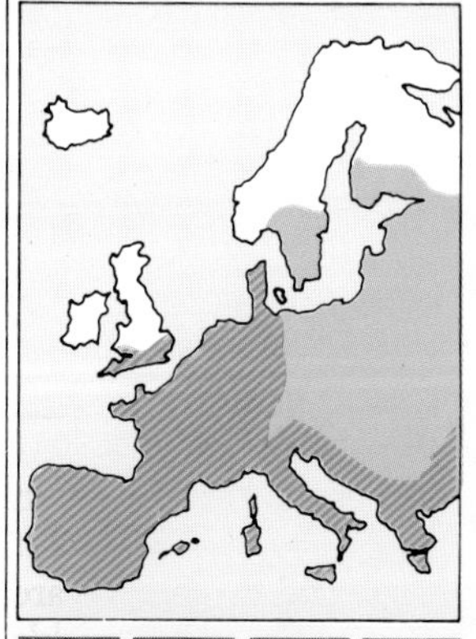

lu-lu-lu

Song A melodious warbling with a slow, rich, descending *lu-lu-lu* phrase.
Behaviour Fluttering song flight, spiralling up then circling before descent.
Habitat Scrubby heaths and hillsides with scattered trees and bushes.
Nest Of grass in a sheltered depression on the ground. Three or four eggs, two broods.
Food Beetles, flies and moths, some seeds in autumn and winter.

SKYLARK *Alauda arvensis*

THE TIRELESS SINGING of the skylark has inspired the admiration of poets and musicians for centuries. A bird of open fields, it has mastered the use of the song flight better than any other and seems to extol its joyous freedom with every note. Although the song is particularly associated with spring, skylarks sing all year, except for August and September, when they moult. This is one of the few species that has benefited from the removal of woodland and the cultivation of land for arable farming.

On close inspection, the skylark is a rather dull, brown bird. It spends most of its time on the ground; like all larks, it runs well, and crouches if alarmed. In flight the distinctive white trailing edge to the wing and white outer tail feathers are visible. The crest is most noticeable when raised to show excitement. This bird tends to flutter for short distances, but can fly strongly, with an undulating, if rather floppy, flight.

The song is a rich and varied warbling which may last two to over five minutes. The male begins to sing as he flies upwards, continuing until he is high in the sky, where he may remain for several minutes. When disturbed from the ground he utters a *chirrup* call, which he also gives when flying in a flock.

The nest is built on the ground in grass or crops. When returning to the nest, the adults always land some distance away and then quietly run unseen through the grass to reach it. Three or four eggs are laid, and the pair may raise two or three broods.

Skylarks are resident in lowland areas but tend to leave higher regions when the weather becomes inhospitable in winter. They often gather in winter feeding flocks of several hundred birds; they roost on the ground in long grass, sheltering in a depression from the wind. In October and November large numbers of skylarks migrate from northern Europe to spend the winter farther south. Many come to Britain, and when the winter is particularly severe, British skylarks may leave for the Continent. Their diet is a mixture of seeds, worms and insects. In the autumn they eat cereal grains, relying more on weed seeds later in the winter.

FACTS AND FEATURES

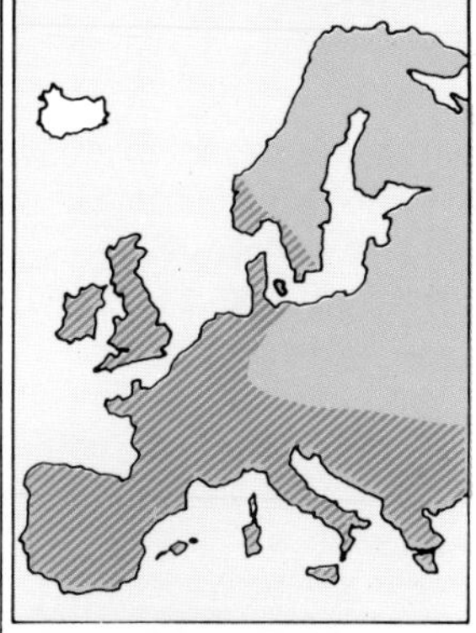

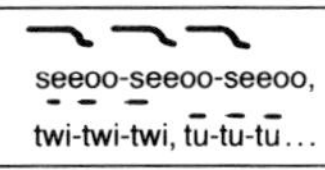

Song A rich and varied warbling with whistling phrases. Lasts for five minutes or more, longer than other larks.
Behaviour Sings in song flight, starting as it ascends, continuing until it is hanging high in the sky.
Habitat Open fields, marshes and dunes.
Nest Of grass in a depression on the ground among grass or crops. Three or four eggs, two or three broods.
Food Seeds, worms and insects. Mainly grains in autumn and weed seeds in winter.

The song in flight of the skylark is the richest and longest of all the larks. In flight the skylark shows white outer tail feathers and a white trailing edge to the wing. One of the largest larks, it has dull brown plumage and a slight crest that is noticeable when raised.

SAND MARTIN *Riparia riparia*

FACTS AND FEATURES

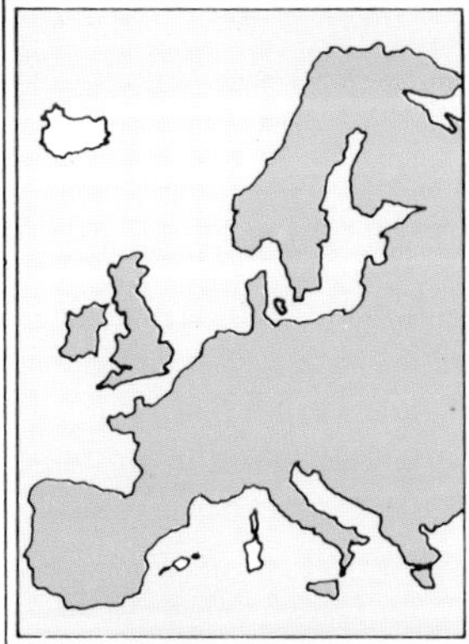

dzzrrr, dzzrrr, dzzrrr

Song A buzzing, chattering twitter.
Behaviour Sings on the wing and when perched on a branch or at the nest hole.
Habitat Open country near water.
Nest Of grass and feathers at the end of a horizontal burrow about three feet (one metre) long. Four or five eggs, two broods.
Food Insects caught on the wing, mainly flies and beetles.

ONE OF THE FIRST BIRDS to return in spring, from its wintering grounds far to the south, is the sand martin. It is small bird and its short tail with a shallow fork, white throat with a brown breast band, brown back and erratic, flitting flight all distinguish it from other European martins. After spending winter in the Sahel region of Africa, south of the Sahara, sand martins return faithfully to their regular nest sites several thousand miles to the north. In recent years there has been a decline in the numbers of sand martins returning, with the British population falling by nearly 95 per cent. This crash has been attributed to the drought conditions in the wintering areas, which have meant that fewer birds have survived the winter, and those that have are less able to make the journey north.

A sand martin colony is fascinating to watch as the birds fly back and forth to their nest burrows, chattering incessantly. The song is not exactly musical, but more of a dry, buzzing twitter delivered both on the wing and when perched. The call is a harsh *chirrup* given from the nest hole and in flight.

The nest is built at the end of a horizontal tunnel in the face of a sand quarry, sandy cliff or river bank. It is possible to encourage these birds to breed by digging holes in a sandy bank and inserting lengths of clay pipe. This encourages them to inspect the holes and continue excavation. They lay four or five eggs and often have two broods.

Sand martins feed on insects caught in flight, usually over a nearby river or lake. These summer visitors to Europe arrive in March; after breeding they depart as early as July, with some staying to September or later. They roost together in reedbeds, often with swallows, and make plenty of noise when gathering at dusk. Like other members of their family they collect in large parties of adults and young before migrating south.

The superficially similar crag martin, *Ptyononprogne rupestris*, is larger than the sand martin and has brown underparts. It is found on southern European cliffs and mountains and has a weak, twittering song.

A small member of the swallow family, the sand martin is most easily recognized in flight by its brown colour, with white underparts and a brown collar, and its short tail, which is only slightly forked. The sand martin is rarely to be found far from the water where it feeds, and from the sand and gravel areas where it excavates its nest.

SWALLOW *Hirundo rustica*

ONE OF THE TRADITIONAL SIGNS of spring, together with the sound of the first cuckoo, is the return of the swallow. These birds can be seen arriving along the south coast of Britain in April, flying in low over the water and almost immediately starting to hawk for insects as they reach land. Their buff-white underparts, chestnut throat and two long tail streamers make them immediately recognizable. Swallows are more graceful in flight than other members of their family, swooping and turning as they chase insect prey. They are found in open country, especially near water, and can be seen feeding over meadows, lakes and rivers.

The swallow sings both in flight and when perched. Telephone wires are a favourite spot, but fences and branches are used as well. The song is a twittering warble which ends with a trilling *twee-ee*. Singing commences as soon as the birds arrive and continues regularly until July, being heard sporadically until September. The call on the wing is a *tswit, tswit, tswit* and *tse-tswit*. If danger threatens, a high *tswee* alerts other birds.

Swallows have taken advantage of human presence, using barns and outhouses as nest sites and returning to them every year. The nest is built of grass and mud, the latter collected from a nearby river's edge or farm pond. It is sited on a ledge or beam in a shed or porch; one pair nested on the curtain rail of a bedroom! Swallows lay four or five eggs and have two broods.

This species travels farther than almost any other European passerine, flying all the way to South Africa for the winter. Many do not complete the 6,000-mile (10,000-kilometre) journey, especially if they are intercepted by hunters in the Mediterranean. The first swallows reach southern Europe in early March, arriving in Britain in April and in northern Europe by early June. In autumn, birds leave Britain in late August, but a few stragglers can still be seen in November. They feed mainly on small flies but take other insects such as beetles and flying ants.

A similar species, the red-rumped swallow, *Hirundo daurica*, breeds in southern Europe. It has shorter tail streamers and an orange-buff rump compared with the swallow; its song is similar, but shorter and slower.

In summer the swallow is easily identified in flight by its long tail streamers. At close range the white spots on the tail feathers can be picked out. Although the young birds have shorter tails, they have the dark back, pale underparts and the orange-buff throat of the adult swallow. The swallow spends more time perched on wires than others of its family and will gather with sand and house martins before migrating in the late summer or early autumn.

Having left the nest, these fledglings (right) *are anxiously waiting for an adult to return with food. They will stay together for some time and often return to the nest at night to roost.*

Swallows and house martins often congregate in late summer forming large flocks prior to migration. Most of these are young birds and telegraph wires (above) *are a favourite perch for preening.*

These young swallows (left) *are about to leave the nest; note the pale yellowish gape around the bill. The adult is keeping its balance by half perching and flapping its wings.*

FACTS AND FEATURES

weetaweet, weetaweet,
weetaweetit, twee-ee

Song A twittering warble with a trilling *twee-ee* at the end.
Behaviour Sings when perched on a wire or branch and also in flight.
Habitat Open country near water.
Nest Of mud and grass, on a ledge or beam in a barn, shed or porch. Four or five eggs, two broods.
Food Flying insects, mainly flies, beetles and ants.

HOUSE MARTIN *Delichon urbica*

Aptly named from its habit of building mud nests under the eaves of houses, the house martin's distinguishing features are its blue-black upperparts with a white rump and all-white underparts. It lacks the tail streamers of a swallow and has a bulkier build than a sand martin.

A FAMILIAR BIRD in both town and country, the black-and-white house martin is well named since it rarely nests anywhere other than on or against a building. Regarded by many as an omen of good luck, this bird creates some ill will because of the mess that its nesting can make; it always seems to pick an inconvenient place, so that droppings land on a doorway or window ledge.

House martins are masters of the air, feeding on the wing. They are rather fluttering, but can glide well and are exceptionally agile, wheeling around when chasing flying insects. On the ground they are very ungainly and tend to shuffle along, only landing to collect mud for the nest. Their legs are unusual in that they are completely feathered – even the toes. Closely related to the swallow, the house martin is easily distinguished by its completely white underparts, white rump and less forked tail.

The song is a rather unmusical twittering, given in flight and also from the nest. These birds can be noisy, and a colony of nests, full of young, can make quite a din that is heard indoors if they are positioned close to a window. Their call note is a hard *chirrup*.

House martins build mud nests and their favourite position is under the eaves of a house. Some still use traditional cliff nest sites, but these are now rare. The nest has a small entrance hole at the top which keeps out sparrows – unless these usurpers take over before the nest is completed. Birds return to the same sites each year. They are colonial and often many nests are built side by side. This tendency helps when trying to attract them to breed, using an artificial mud nest fixed under the eaves. The artificial version may induce birds to build alongside it, even if they do not use the nest itself. Two broods are normally raised and the young from the first often help the parents to feed the next brood.

House martins spend the winter in tropical Africa. They begin to arrive in southern Europe as early as February and reach Britain in early April. After an initial inspection of their nests they often disappear for some days to feed, before refurbishing the nest and breeding. They feed on flying insects, especially flies and beetles and sometimes butterflies and moths.

FACTS AND FEATURES

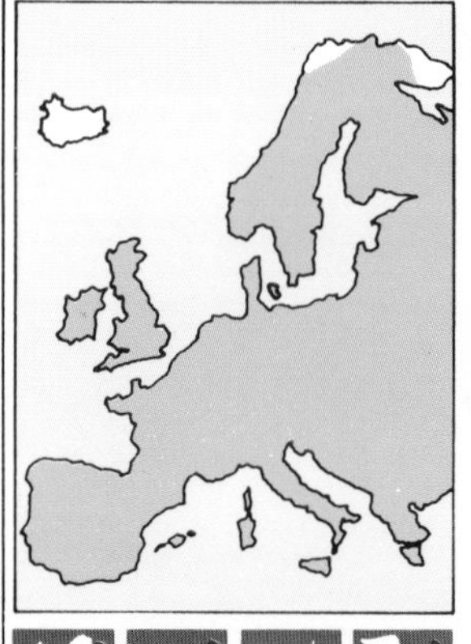

prritt, prritt, prritt

Song An unmusical twittering.
Behaviour Sings in flight and also when perched, usually at nest.
Habitat Open country around towns and buildings.
Nest Of mud collected from nearby pond or river bank. Built under eaves of building, or occasionally on cliffs. Four or five eggs, two or three broods.
Food Insects caught in flight, mainly flies and beetles, also butterflies and moths.

ROCK PIPIT *Anthus petrosus*

THE ROCK PIPIT is an inconspicuous bird and its presence is usually only betrayed by its high, squeaking call, frequently given in flight. Almost exclusively coastal, rock pipits spend a great deal of their time on the shore. Here they search for food, running in and out of rocks and pecking among the seaweed – behaviour which has earned them the local name of 'sea mouse'. They habitually perch on rocks, walls or other prominent features. They are dark and stouter in appearance than their cousins the meadow pipits. The dark olive upperparts, buff underparts with poorly defined breast streaks, dark legs and grey-white outer tail feathers distinguish them. They are usually solitary, occasionally forming flocks in winter.

The rock pipit has a loud song which starts like a series of call notes, accelerating and becoming more musical before ending with a trill. The song is usually delivered in flight, the bird rising from the ground while beginning to sing and then descending while giving the main song. The call is similar to a meadow pipit's but fuller and louder – a *weest*, uttered singly and repeated less often compared to a meadow pipit. The song is given from late March to early July and is not heard outside this breeding season.

The nest is usually in a rock crevice, never far from the coast, and generally in a sheltered gully. The female lays four or five eggs and the pair often raises two broods.

Resident over much of their breeding range, birds move along coasts in winter, when they can be seen on estuaries and saltmarshes. They feed on insects, small worms, crustaceans, molluscs and seeds. In coastal areas they visit bird tables for winter food and might breed in the stone wall of a cliff-top garden. Northern populations migrate south in winter and the Scandinavian race, *A. p. littoralis*, can be found along the east and south coasts of England, as well as on the shores of the Netherlands, Belgium and France.

The rock pipit was once considered to be a sub-species of the water pipit, but recent changes have given it full species status. The two birds are similar but do not share the same habitat.

FACTS AND FEATURES

see-see-see-see-
sui-sui-sui-sui-sweee

Song A series of squeaky notes accelerating and ending with a trill.
Behaviour Sings in display flight with bird ascending and descending to a perch. Sometimes sings from ground.
Habitat Rocky coasts, shores and estuaries.
Nest Of grass in a rocky crevice, close to the coast. Four or five eggs, two broods normally.
Food Feeds on insects, worms, crustaceans, molluscs and seeds along shoreline.

The most inconspicuous of all the pipits, the rock pipit is larger and bulkier than tree or meadow pipits, and has darker legs. Its back and breast streaks are poorly defined and in flight it shows grey-white outer tail feathers. The rock pipit's liking for rocky coastal areas will also help to distinguish it from its inland-based cousins.

WATER PIPIT *Anthus spinoletta*

THE WATER PIPIT, one of Europe's most attractive pipits, breeds in the more mountainous regions of the Continent. It nests in alpine pastures, on rocky slopes and stream edges at altitudes of between 5,000 and 8,000 feet (1,500 and 2,500 metres).

Unlike other pipits, this species has an almost unstreaked brown back, greyish head and whitish underparts, which have a pink flush in the breeding season. It also has a bold, pale stripe over the eye; like the rock pipit, it has dark legs. In winter there is some streaking on the underparts, but the bird retains the eyestripe and white outer tail feathers, which helps to distinguish it from the rock pipit. As the name implies, at all times of the year the water pipit tends to be found near water. The close relationship of this bird to the wagtail can be seen in the way the former runs along the ground, slightly wagging its tail.

In the breeding season water pipits perch on rocks and bushes to sing, and also indulge in a song flight. The song is similar to that of the tree pipit, but does not have such long, drawn-out notes towards the end. The call is like a rock pipit's, but is perhaps a little less shrill, often described as *tseeep*.

FACTS AND FEATURES

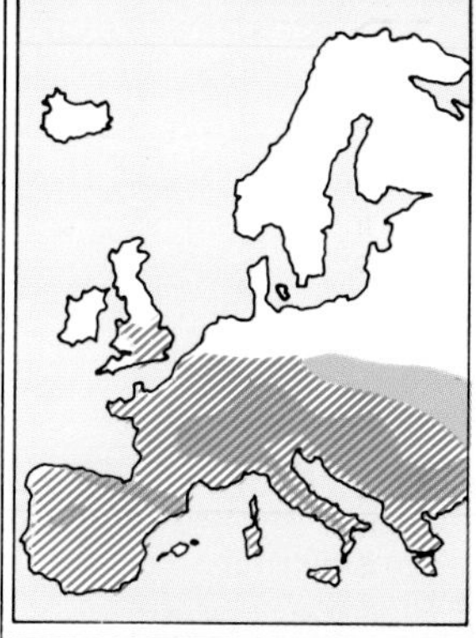

see-see-see-see-see-se
see-see-see

Song A short phrase of repeated high notes, slowly accelerating.
Behaviour Sings briefly from rocks or bushes, and for longer in song flight.
Habitat Mountainous areas and inland waters.
Nest Of grass and concealed in a bank or grassy hollow. Four to six eggs, two broods.
Food Beetles, flies and their grubs, spiders, small molluscs and seeds.

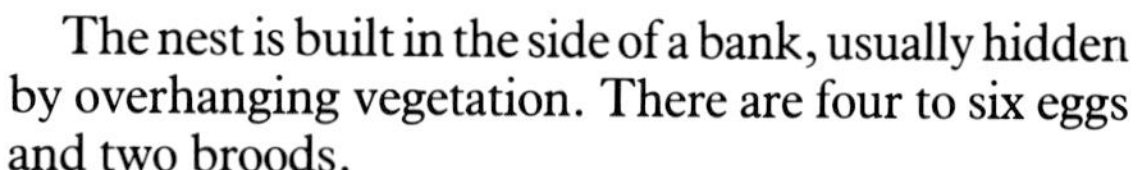

The nest is built in the side of a bank, usually hidden by overhanging vegetation. There are four to six eggs and two broods.

In the autumn, water pipits leave the high ground and move to lowland areas for the winter. Some stay on the lower, grassy slopes of the hills where they bred; others move to wet grasslands and marshy areas nearby. Small numbers, probably young birds, wander to Britain each year. They favour sewage farms, watercress beds and lakesides, where they walk daintily along the fringes in search of food.

The true status of the water pipit has been argued over for many years – whether it deserved to be a full species, or included as a sub-species within the same species as the rock pipit. While these two birds are similar vocally, the differences in habitat preference and distribution have resulted in their recently adopted position as two individual species.

The water pipit is easily differentiated from the rock pipit in spring, when its breast is unstreaked and the bold, pale stripe over the eye is obvious. In flight it shows white outer tail feathers. Its affinity to water, especially in the breeding season, will usually help to identify it.

TREE PIPIT *Anthus trivialis*

FACTS AND FEATURES

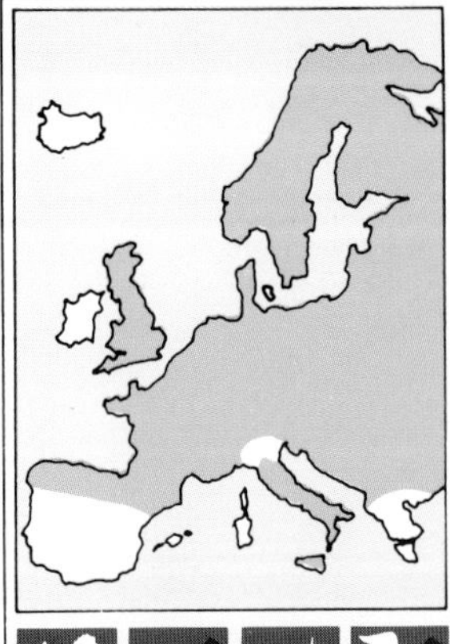

...seeea-seeea-seeea-seeea

Song A powerful song with short phrases repeated before finishing with a few drawn-out, more musical notes.
Behaviour A 'parachute' display accompanies the song as the bird floats down on spread wings and tail.
Habitat Trees, bushes, meadows, heaths and hillsides.
Nest Of grass and moss, hidden on the ground among grass. Four to six eggs, possibly two broods.
Food Beetles, grasshoppers, flies and their grubs, and spiders. Some seeds in the winter.

A SUMMER VISITOR to Europe, the tree pipit's flamboyant song flight and vocal outbursts make it one of the most notable arrivals in spring, next to the nightingale. Open areas of heathland, scrub with scattered trees, and young plantations form its usual home. As the name suggests, this bird prefers areas with trees and bushes, from where it launches itself into song.

The plumage is like that of most pipits – brown-streaked upperparts, pale underparts with streaks on the breast and flanks, and white outer tail feathers. The tree pipit has a slim body and pale legs, running along the ground to feed but habitually flying into trees when disturbed. At close range it can be seen to have shorter hind claws than the meadow pipit, indicating it spends less time on the ground.

Males arrive first at the breeding ground and take up their territories. The song flight begins from a perch, usually a tree branch or the top of a bush. The male flies upwards, fluttering as high as 100 feet (30 metres); as he reaches the highest point, he begins to sing and then floats down to the ground with wings and tail spread out. The main song, given during the descent, is a repeated phrase of a few notes leading up to a powerful *seeea-seeea-seeea*... before landing on a perch. It is the most far-carrying song of all the pipits and is heard from early April until July. The call, usually given in flight, is a single and rather hoarse-sounding *teeez*.

Tree pipits nest on the ground in a grass tussock or bank, and lay between four and six eggs. Birds that arrive in early spring may have enough time to raise two broods, although many manage only one, especially in northern Europe. They feed on insects taken on the ground, but also eat seeds in winter.

The first spring visitors arrive in Europe in early April; most have reached northern Europe by the end of May. They leave for their winter quarters, the grassy savannas of tropical Africa, from late July to October. On migration they are usually seen singly, unlike meadow pipits, which are generally in small groups.

A brightly coloured member of the pipit family, the tree pipit has black upperparts streaked olive-brown, lightly streaked yellow-buff breast, white belly, white outer tail feathers, and pink legs. It habitually perches in trees and bushes but feeds on the ground. Its loud, rather hoarse teeez *call from April to July identifies it in flight.*

MEADOW PIPIT *Anthus pratensis*

'MOOR PEEP' IS ONE of the many descriptive local names for the meadow pipit. It is a small brown bird which breeds and winters in a variety of habitats. It closely resembles the tree pipit, having a brown-streaked back and streaked breast.

Meadow pipits feed in open country among rough grass and heathland, and often their presence is only noticed as they fly up, calling with a thin, squeaky note. If a feeding flock is disturbed the members usually fly away one by one, jerkily ascending. On the ground they are active birds, running between clumps of grass and pausing occasionally with tail wagging slightly. They sometimes perch on clumps of heather, but rarely fly up into trees.

These birds breed on moorland and rough pasture. The nest is built in a tussock of grass or heather and four or five eggs are laid, with two broods usually reared. Meadow pipits are commonly hosts to the cuckoo, which can produce eggs that closely mimic their own. They leave their highland territories in winter for the more hospitable lowlands. Upland birds are a favourite prey of the merlin, while wintering birds on pastures, marshes and sea coasts are often caught by hen harriers.

The song of the meadow pipit shares similarities with other pipits, as does the song flight. The male starts to fly up, giving a series of thin notes which resemble the call. These notes gather speed as the bird ascends; when he floats down to earth, they change to a rapid succession with a trill at the end. Singing begins in March and continues to July, sometimes later. The call note is a repeated thin *tseep*.

FACTS AND FEATURES

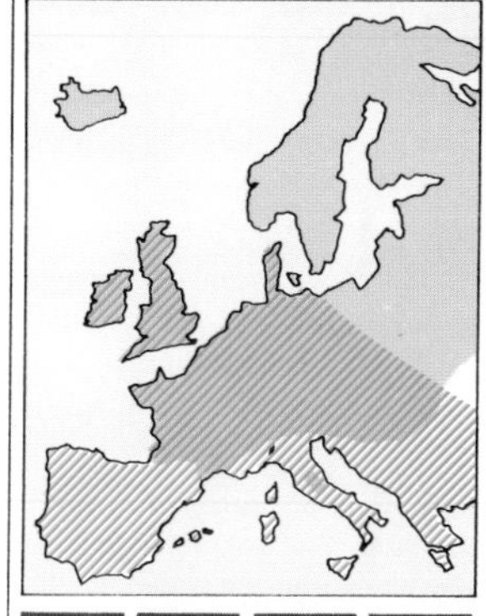

twi-twi-twi-twi fwee, fwee, fwee, fwee, trrrrr

Song A series of thin notes which increase in speed and end with a trill.
Behaviour Full song is given in fluttering song flight; a shorter version is delivered when perched.
Habitat Grassland, moors, fields, marshes and coasts.
Nest Of grass in a depression in the ground among heather and grass. Four or five eggs, two broods.
Food Beetles, crickets, flies, bugs, worms, spiders and occasionally seeds.

Meadow pipits are common breeding birds throughout northern Europe, but only occur as winter visitors to the south. Scandinavian birds migrate south for the winter and many of them pass through or stay in Britain. They can be found wintering in suburban areas, feeding on open grassy areas and sometimes entering gardens. They eat insects, earthworms and occasionally seeds.

A slim, streaked pipit, the meadow pipit is similar to the tree pipit, but its underparts are streaked more densely. It has a squeaky call – a reedy-sounding tseep *repeated several times. The meadow pipit is often seen in small flocks.*

Tawny Pipit *Anthus campestris*

A large, pale bird with long legs, the tawny pipit has a long tail which gives it a wagtail-like appearance. It also has an unstreaked breast and sandy-coloured back in breeding plumage. (Young birds have a streaked back and some streaks on the upper breast.) The buff stripe over the eye is always noticeable.

THIS SPECIES IS a large, slim and pale pipit with long legs, that spends most of its time on the ground and runs rapidly. It has the slim appearance and horizontal posture of a wagtail, accentuated by its long tail; unstreaked underparts in summer distinguish it from other pipits in its dry and open habitat. It is often located by its song and calls, yet can be very difficult to see as it crouches behind a tussock of grass or runs among stones. If it perches on a rock its slim shape, long legs, pale eyestripes and dark moustache stripe are diagnostic features. In autumn, young birds have a streaked breast.

The tawny pipit's song consists of a series of clear, metallic notes given during a rather fluttering song flight. It resembles *chiveee, chiveee, chiveee*... and is sung during the descent. The bird sings from April to July. The call notes are quite distinctive: when disturbed or alarmed it gives a soft, sparrow-like *chirrup*, and in flight it makes a *sweep* like the yellow wagtail.

The nest is made in a depression on the ground, well hidden amongst tussocks of grass. Like so many ground-nesting birds, this pipit lands away from the nest and creeps there unobserved. Four or five eggs are laid, and two broods may be reared in a good year.

Tawny pipits spend the winter in Africa, north of the equator, and begin arriving in their breeding areas during March and April. They return south in August. These birds occur as rare visitors to Britain, both in spring and autumn, usually on the south and east coasts. They feed on insects and their grubs, taken on the ground.

The similar Richard's pipit, *Anthus novaeseelandiae*, is an uncommon visitor to Europe in the autumn. It has a streaked breast (like a young tawny pipit) and its call is a harsh *schreep*.

FACTS AND FEATURES

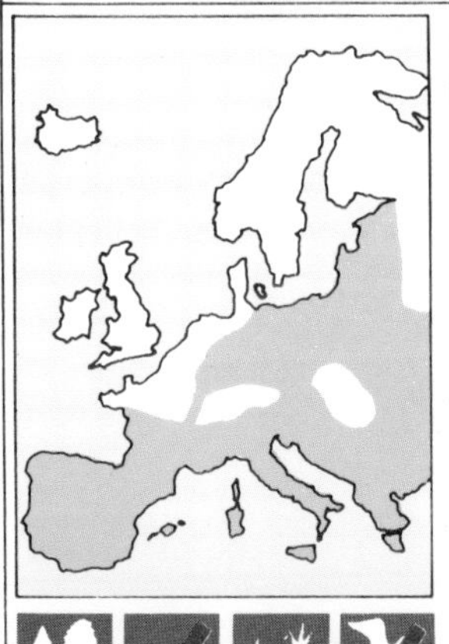

chiveee, chiveee, chiveee

Song A loud and monotonous series of notes, *chiveee, chiveee, chiveee*.
Behaviour Song given in a fluttering display flight, during descent.
Habitat Open areas with scrub.
Nest Of grasses and roots, well hidden in a depression in the ground among grass tussocks. Four or five eggs, two broods.
Food Beetles, crickets, flies and their grubs.

YELLOW WAGTAIL *Motacilla flava*

THE ARRIVAL OF THE FIRST yellow wagtails in spring is always a great delight. The bright yellow males appear first and can often be seen in flocks, feeding in a waterside meadow, where they outshine even the marsh marigolds. They are very active birds, always running around busily in search of food. They associate particularly with cattle, which disturb insects for them to catch. They are normally found in damp meadows, pastures and fields near lakes and streams, but are not restricted to areas near water. Females are less striking than males, being duller and paler.

In flight these birds have the typical 'bounding' action of all wagtails, which is especially pronounced as they come in to land. Outside the breeding season they tend to form small flocks and in autumn groups can be seen flying south, calling as they go.

The song consists of short, warbling *tsip-tsip-tsipsi* phrases broken up with notes resembling the call. It is delivered from a perch or in flight, but there is no accompanying display or song flight. The call note is a loud and distinctive *twseep* given both in flight and on the ground.

The yellow wagtail's nest is built in a depression on the ground among thick vegetation and six eggs are usually laid. In southern Britain the species usually has two broods, but farther north only one may be raised.

Yellow wagtails return from their African winter quarters in April. They breed over most of England and Wales, although they are scarce in the west. They are absent from Ireland, except as a rare migrant, and breed in only parts of southern Scotland. They feed on insects, especially flies, and some small molluscs, departing in September and October. There are a few genuine records of birds over-wintering in England, but care must be taken not to confuse them with the more resident grey wagtails.

The yellow wagtail is a sub-species of the blue-headed wagtail, which breeds on the Continent and is an occasional visitor to Britain. Male blue-headed wagtail have a blue-grey crown, white chin and white supercilium ('eyebrow'). Other sub-species that breed in Europe include grey-headed, ashy-headed, black-headed and Sykes's wagtails.

FACTS AND FEATURES

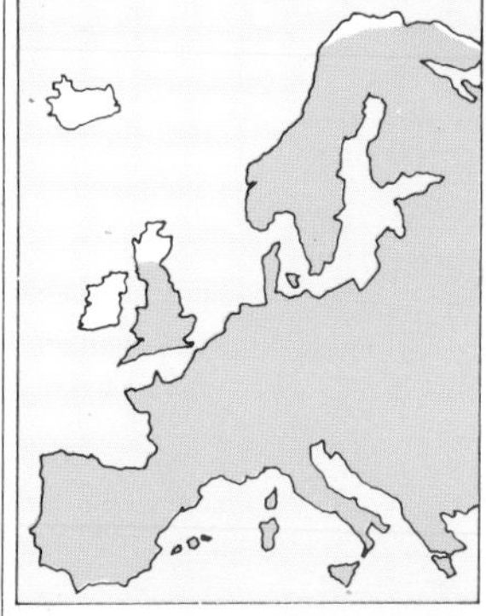

Song Fairly quiet and simple with short, warbling phrases, *tsip-tsip-tsipsi*.
Behaviour Sings from a branch or tussock, and in flight but without display.
Habitat Meadows, usually near water, and marshes.
Nest Of grass and roots on the ground among thick vegetation sometimes crops. Six eggs, two broods.
Food Various flies, and snails and other small molluscs.

In spring the male is unmistakable with its bright yellow underparts and greenish back. Different races have characteristic head patterns. The female is duller-looking with brownish upperparts and paler yellow underparts. In flight its white-edged black tail is evident. Young birds have a pale yellow-buff belly and black throat markings.

GREY WAGTAIL *Motacilla cinerea*

FACTS AND FEATURES

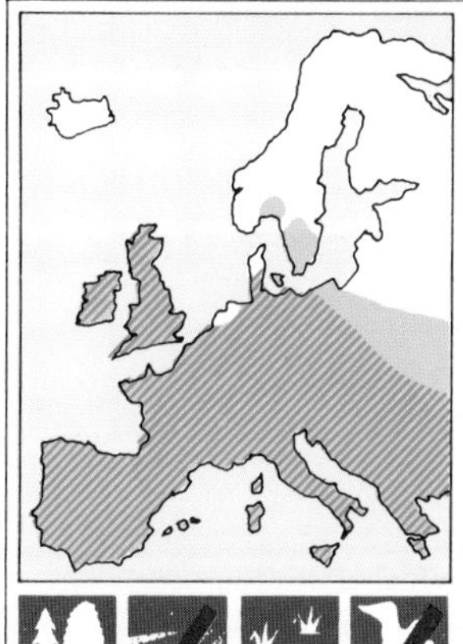

sisisisi, tswee-tswee-tswee,
churrr, trrrr

Song An infrequent mixture of musical and trilling phrases.
Behaviour Sings from rocks and also in a display flight with flickering wings.
Habitat Rocky streamsides, lakes and rivers.
Nest Of moss, leaves, twigs and grass in a bank hole, among rocks or on a ledge. Four to six eggs, one or two broods.
Food Flies, mayflies, water beetles and damselflies.

THE GREY WAGTAIL is one of the most attractive members of its family, and rather unfairly named since only its back is grey – its underparts are bright yellow, often causing it to be mistaken for a yellow wagtail. It has the longest tail of all its European relatives, at half its body length, which means that its habitual tail-wagging is very accentuated. This graceful bird trots along the stony edge of a river, breaking into a run as it darts after an insect, then pausing and vigorously wagging its tail. It may sit on a stone in the middle of a stream and suddenly fly out to catch a passing insect. The flight is strongly undulating and again the tail length helps to distinguish it from other wagtails when on the wing.

This species' song has many of the qualities of a pied wagtail's. It begins with notes similar to its call, breaking into more musical phrases using some short, wren-like trills. The trills are sung particularly in a slow display flight as the bird goes from perch to perch, wings flickering and tail spread. It sings infrequently from late March to May, and occasionally in the autumn. The call also resembles a pied wagtail's, but is a shorter and sharper *tsipp* or *tzitzi*, usually when in flight.

In the breeding season, grey wagtails are found in hilly or mountainous regions near fast-flowing streams and rivers. The nest is built close to running water, usually on a ledge or in a hole among rocks and tree roots. Sometimes the old nest of another bird, such as a dipper, blackbird or swallow, is used. There are four to six eggs, and one or two broods may be reared.

Grey wagtails are mostly residents, but in autumn they move from uplands to lowland wintering areas. They spend the cold months by rivers, lakes, farmyards, watercress beds, coastal marshes, gardens and even around buildings in urban areas. Birds from northern Europe migrate south and some British individuals may go to France and Spain for the winter. In the breeding season they feed on insects and aquatic invertebrates, while in winter their diet becomes more varied and includes worms, small fish and, in severe weather, household scraps. Prolonged snow and frost affect this species drastically, with large numbers dying from lack of food.

A lively long-tailed member of the family, the grey wagtail has bright yellow undertail coverts and a greenish-yellow rump. In summer males have a black throat and both sexes have paler yellow flanks. Back, crown and cheeks are grey with a white stripe over the eye. Its long tail (the longest of all the wagtails) is black with white outer feathers, and its habitual tail-wagging is very noticeable.

PIED WAGTAIL *Motacilla alba*

The black-and-white face pattern of the pied wagtail is very distinctive. The black crown and the white cheeks and forehead contrast with the black throat in summer and the white throat with its black bib in winter. In Britain the back colour is black, but on the Continent it is pale grey. Young birds often have a yellowish tinge to the white parts. The ever-wagging tail is black with a white edge.

THIS SMALL BLACK-AND-WHITE BIRD, with its ever-active tail, is well named. All wagtails wave their long tails up and down incessantly and move their heads backwards and forwards as they walk. This species is the most conspicuous member of its family, found in open country, gardens, parks and farmyards. It can run surprisingly fast as it darts and dashes after insects. In paddocks and fields it may feed around the feet of horses and cattle, snapping up the insects they disturb. The flight is distinctly undulating, becoming more marked as the bird swoops to land.

Pied wagtails collect in small groups at dusk near a communal roost, which could equally well be a remote reedbed or a bush or tree in a city centre. Roosting flocks can number several hundred.

Although closely related to pipits, wagtails differ in not having a song flight. The pied wagtail's song is a simple warbling twitter incorporating slurred call notes. It can be given in flight, but is more usually heard from a perch. In spring, roosting parties may indulge in excited choruses. The common call note is a sharp *chizzick* and there is also a territorial note, a short *chee-wee*, given from a perch.

Holes in walls, banks, thatch and ivy, and ledges in sheds, are all used as nest sites. These wagtails may inhabit an open-fronted nestbox suitably concealed in ivy or recessed into a wall. Nesting starts in April and pairs often fall prey to the parasitic habits of cuckoos. They have two, and sometimes three, broods.

Pied wagtails are resident in Britain, but they forsake higher ground in winter for the milder lowland areas along rivers and lakes, and even town parks. Some birds may fly to the Continent, especially during hard winters. The main food consists of insects, and because of this, they often suffer high mortality if there is a prolonged spell of severe weather.

This species is found over all of Europe, from Iceland to Greece. However, only in Britain is it known as the pied wagtail; there is a different sub-species, known as the white wagtail, elsewhere in Europe. White wagtails differ from pied in their back colour: in the breeding season it is pale grey, rather than black. Female and winter-plumaged pied wagtails can have a greyish back, but never as pale as the white wagtail's. The northern and central European populations are migratory, wintering in the Mediterranean and Africa.

FACTS AND FEATURES

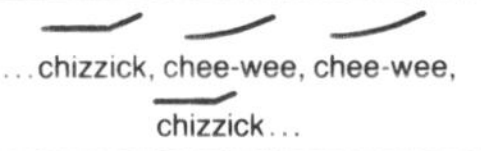

Song A lively, twittering warble incorporating several *chizzick* call notes.
Behaviour Song is usually given when perched, occasionally in flight.
Habitat Fields and farmland near ponds and rivers.
Nest Of moss, twigs, leaves and roots in holes in walls, banks, thatch, on ledges in sheds and in open-fronted nestboxes. Five or six eggs, two or three broods.
Food Flies, beetles, small moths, fish fry, grains and seeds.

DIPPER *Cinclus cinclus*

THE DIPPER is a small thrush-like bird that frequents rivers and streams. It can usually be seen perched on a rock in the middle of fast-running water, or flying with fast wing-beats low over the surface. It is plump and fairly short-tailed with a blackish-brown back, and dark underparts which contrast with a white throat and upper belly. It runs quickly across rocks and when stationary it often bobs up and down, jerking its tail before plunging into the water.

The dipper is remarkable in that it feeds underwater, swimming through the rapid currents and also walking along the stream bed in search of food. It eats aquatic insects and their larvae, molluscs, crustaceans, worms and occasionally small fish. This bird uses the current of the water to help keep itself submerged, by facing upstream and tilting its tail up into the current; it bobs out of the water and onto a rock with ease.

The dipper's song is a grating warble given by both sexes, usually from a rock or low branch but sometimes in flight. It has a rippling quality with some trills and can be heard all year, except from July to September. The song is also given during courtship display, when both birds bow at one another and shiver their wings. Calls are a loud *zit-zit-zit* and a sharp *clink, clink*.

Dippers build domed nests close to the water: in a rock crevice, under a bridge, in tree roots or sometimes on a branch overhanging the water. Five eggs are usually laid and two broods, occasionally three, are reared. Each pair occupies a distinct territory along a river and rarely leaves it.

FACTS AND FEATURES

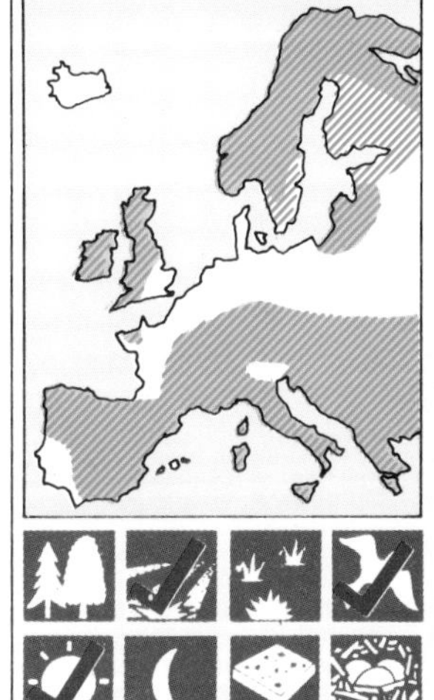

whee, ti-ti, wheeoo, ti...

Song A rippling, slightly grating warble with trills, sung by male and female.
Behaviour Sings when perched on a rock or low branch, sometimes in flight. Also given during wing-shivering display.
Habitat Fast streams and rivers with rocky edges.
Nest Roofed nest of moss and grass close to water, in a hole in a wall, among tree roots or in a rock crevice. Five eggs, two broods.
Food Larvae of aquatic insects – beetles, mayflies and dragonflies – and molluscs, worms, tadpoles and small fish.

Dippers are resident birds over much of their range, but birds from northern Europe wander in winter. The British race of dipper has a chestnut band across the belly which distinguishes it from the north European race with a black lower belly. There is a separate Irish race that has darker upperparts than the British. Black-bellied dippers are occasionally seen in Britain in winter, usually in eastern counties.

The dipper's white throat and breast contrast with its dark, stout body. Males and females are identical, but young birds are greyer above with mottled grey underparts. Its favourite haunt of a rock in swift-flowing water can often be identified by the marks of its droppings. The dipper feeds underwater – walking along the stream bed or plunging in from a stone.

WREN *Troglodytes troglodytes*

ALMOST EUROPE'S TINIEST BIRD, the wren is one of the commonest, found everywhere from gardens to moorland, woodland and seashore. It is only absent in the far north, where conditions are too inhospitable in winter. This small brown bird is the only representative of its family in the Old World, but in the Americas there are dozens of related species.

This secretive species spends most of its time in dense undergrowth or skulking under bushes. It is adept at searching every nook and cranny for food, creeping around in a mouse-like way and holding its short tail cocked at an angle.

The wren's song cannot be ignored – because of its sheer volume. It seems incredible that such a small bird can produce such loud sounds, which can be heard at distances of half a mile (nearly one kilometre). It consists of a series of rattling, warbling phrases ending with a ringing trill, and lasts for about five seconds. Although wrens are sometimes hidden when they sing, they also use posts and open branches; but they rarely sing from very high off the ground. Occasionally they will sing in flight, especially if disturbed. They can be heard all year round, but tend to go quiet in the autumn, when they moult. Wrens are one of the first birds in the dawn chorus, beginning about 20 minutes before sunrise. Calls vary from an explosive ticking, *tititic*, to a scolding churr, *scurrr*.

The male wren builds a number of nests and one of these is chosen by the female, who then lines it. Favourite sites are in ivy, hedges, tree holes and walls; sheds and other buildings have been used, and nests have even been built in an old coat pocket and in washing on a line. Usually, five or six eggs are laid and two broods raised.

Wrens are resident in Britain and do not move far from their breeding areas in winter. The diet of insects and spiders makes wrens vulnerable and bad winters have seen their populations halved. In cold weather they may roost communally – as many as 50 have been seen entering a nestbox for the night.

FACTS AND FEATURES

sioosioo-trrrrrrr-shishishi-tootootootoo-i

Song A very loud series of warbling, rattling phrases ending with a trill.
Behaviour Sings from low positions on a bush, post or tree, sometimes in flight and occasionally at night.
Habitat Gardens, woodland and moorland.
Nest Domed nest of moss, grass and leaves with a side entrance, in hedges, ivy, holes in trees and walls. Five or six eggs, two broods.
Food Moth caterpillars, flies, beetles, aphids, spiders and seeds.

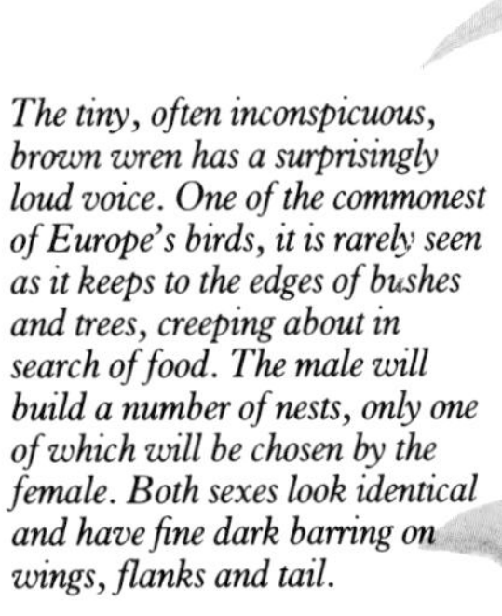

The tiny, often inconspicuous, brown wren has a surprisingly loud voice. One of the commonest of Europe's birds, it is rarely seen as it keeps to the edges of bushes and trees, creeping about in search of food. The male will build a number of nests, only one of which will be chosen by the female. Both sexes look identical and have fine dark barring on wings, flanks and tail.

DUNNOCK *Prunella modularis*

The Dunnock always appears nervous, walking or hopping with its curious shuffling gait, close to cover and ready to disappear if danger threatens. Dunnocks frequently chase one another during the spring, uttering insistent piping calls. Male and female look identical, with grey head and underparts, brown- and black-streaked back and pink legs. The song is similar to that of the wren but more tuneful.

FACTS AND FEATURES

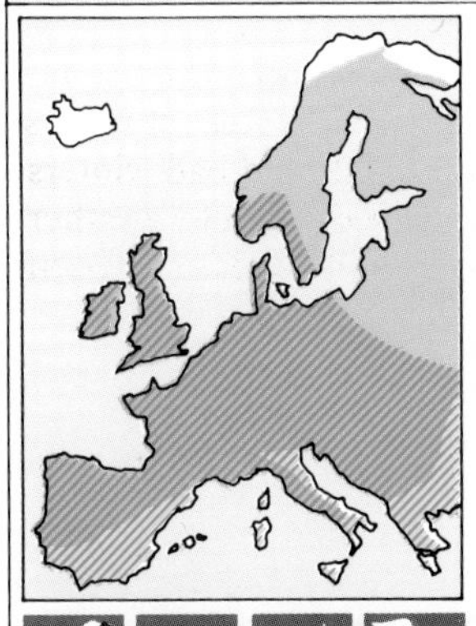

weeso, sissi-weeso, sissi-weeso, sissi-weeso...

Song A high musical warbling.
Behaviour Sings from the top of a hedge, bush or tree.
Habitat Gardens, hedges and bushes.
Nest Of twigs and moss with leaves and roots, in hedges and evergreen bushes. Four or five eggs, two or three broods.
Food Beetles, caterpillars, spiders and worms. Seeds in winter.

THE DUNNOCK, also known as the hedge sparrow, is a common garden bird in most of Europe. It is shy and seldom allows close approach, appearing from a distance as brown with a grey head and belly. Closer inspection reveals a beautifully patterned chestnut-and-black back, reddish eyes and a slim dark bill, quite unlike that of a true sparrow. In fact the dunnock is not a member of the sparrow family, it is an accentor. One of its more colourful local names is 'shufflewing', accurately describing the bird's habits. It spends much time on the ground, walking or hopping with a distinct shuffling gait, flicking its wings as it goes. It is found in woodland, hedges and moorland as well as gardens and parks. This is a solitary bird except when feeding.

The dunnock's song is a high, cheerful warbling, similar to a wren's but more musical. Males usually choose a perch on top of a hedge or bush, or sometimes on a tree branch. Singing occurs throughout the year, but begins in earnest in late January or early February. The call notes are a clear, insistent *tseeep* and a high, rapid trill.

The nest is well hidden in the middle of a hedge or bush, often in a conifer or other evergreen and sometimes in thick scrub. The species is a favourite host of the cuckoo; unusually, the cuckoo is unable to mimic the plain blue eggs of the dunnock, yet the latter appears not to notice the difference. The sight of a parent dunnock feeding a cuckoo fledgling many times its size, perched on its back, can look very comical. Dunnocks have an interesting social life since they do not always form pairs when breeding. A single male or female may have many partners, or a number of males may share several females.

Dunnocks are resident in Britain, rarely straying far from their home territory. On the Continent the northern populations migrate south for winter and some of these are seen along Britain's east coast in the autumn. They eat a varied diet of seeds, insects, worms and other invertebrates.

ROBIN *Erithacus rubecula*

IN BRITAIN, THE ROBIN is a favourite to the extent of being accepted as a national symbol. As with most garden birds, its true habitat is woodland. On the Continent this is where robins are found; there, they are far from tame and confiding, but shy and retiring and rarely seen except on migration.

The robin needs little description. Its other name, redbreast, sums it up in a word. It is difficult not to admire its cheeky nature as this opportunist watches a gardener digging over the earth and hops down to take an exposed worm or other tasty morsel. Robins stand and perch upright and hop rather than walk. If alarmed they bob up and down and raise their tails. They are regular visitors to bird tables, at which they tolerate no others of their kind, aggressively chasing off intruders. Their intensely territorial nature is a feature of their lives, and in song each robin lays eternal claim to the area it regards as its home.

The song is sweet and warbling, with a slightly melancholy tone; it is particularly intense in spring. In autumn a quieter, thinner song is heard. Robins sing throughout the year, except when moulting, from

Garden sheds, garages or disused buildings are often selected as nesting places (right).

Old tin cans and kettles (right) *are readily accepted by robins as suitable places in which to nest. They will also commonly use open-fronted nestboxes.*

FACTS AND FEATURES

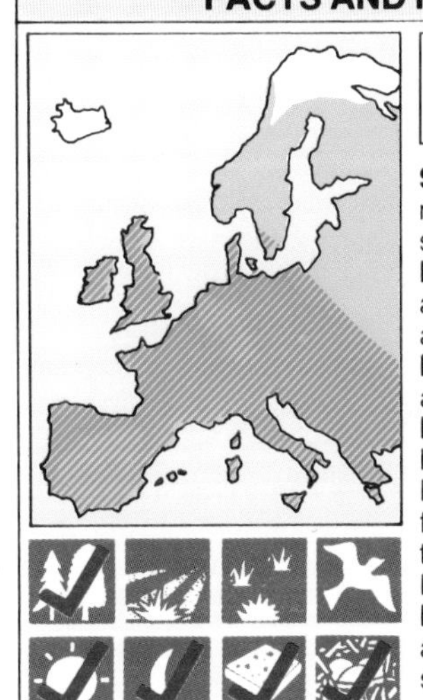

Song A sweet, slightly melancholy warbling with short melodious phrases.
Behaviour Sings from cover and from a perch, sometimes at night.
Habitat Gardens, hedges and woodland.
Nest Of leaves and moss in a hollow in a bank, in ivy, on ledges in sheds and in open-fronted nestboxes. From five to seven eggs, two or three broods.
Food Beetles, caterpillars, ants, flies, spiders, worms, seeds, fruit and berries. Fat, nuts and seeds from a bird table.

cover and an exposed song perch. Reports of nightingales heard during winter are invariably robins, singing at night close to a street lamp or other artificial light. They have a number of calls, the commonest being a sharp, scolding *tic-tic-tic* when excited. Another call is a soft *tseet*. Part of a robin's courtship involves the male feeding the female, stimulated by her insistent *seee-seee-seee*, sounding like a nestling begging for food.

Robins nest in any cavity among vegetation on a low bank, but they readily use man-made nest sites: ledges in sheds and porches, old kettles, flowerpots and boxes. They will occupy an open-fronted nestbox placed in a suitably protected position, safe from cats. They lay between five and seven eggs and may have two or even three broods.

British robins are resident and in the autumn they are joined by their northern European relatives. The migratory nature of robins on the Continent has led to their being shot as they move to the Mediterranean. They eat insects, worms and berries but take almost anything from a bird table in cold weather.

Cheeky and unmistakable, the robin is one of the tamest of Britain's garden birds. Robins are highly territorial, chasing out intruders. In towns, they often nest close to buildings. The males and females are indistinguishable, but young birds can be easily recognized because they lack the red breast of the adult bird and their buff-edged feathers give them a speckled appearance.

NIGHTINGALE *Luscinia megarhynchos*

FEW BIRDS have been as immortalized in verse as the nightingale. Seldom seen, but once heard never forgotten, the loudness of its song makes it easy to pinpoint the bird's location – yet its secretive and skulking nature means that, even when a few feet away, it can still remain invisible. When glimpsed, the nightingale appears a drab brown bird with greyish underparts, only livened by its rufous brown tail.

In Britain, the range of this species has decreased during the last 100 years, as deciduous woodland has disappeared and forestry management practices have changed. It prefers open woodland with a thick layer of ground vegetation (such as nettles and brambles), coppiced woods and scrubby thickets; the advent of large-scale conifer plantations has not suited it.

In common with many dull-coloured birds that inhabit dense cover, the nightingale's song is loud and intricate and its quality is rich, liquid and piping. Phrases within the song are repeated and a loud and rapid *chook-chook-chook*... is one of the most striking. Each burst of song often ends with a fluty *pioo, pioo, pioo*... that rises slowly into a crescendo. Usually hidden when singing, occasionally birds choose a perch on the outside of a bush or a low branch, in full view. Singing begins when the first males arrive in mid-April and continues until mid-June, when the eggs have hatched. The nest is built on or close to the ground, in grass or nettles deep within a thicket.

This is the only songbird in Britain that sings habitually at night as well as during the day. Its commonest call note is a soft *hweet* similar to a loud chiffchaff, and it also gives a hard *tac-tac* and a harsh *krrr* of alarm.

FACTS AND FEATURES

chook-chook-chook... pioo, pioo, pioo

Song A loud, rich and liquid song with piping and fluty notes. Common phrases are *chook-chook-chook* and *pioo, pioo, pioo*.
Behaviour Highly secretive, singing from dense cover, occasionally from the edge of a bush or tree.
Habitat Woodland, thickets and bushes.
Nest Of leaves and grasses on or close to the ground in thick vegetation. Four or five eggs, one brood.
Food Beetles, caterpillars, flies, spiders and worms as well as fruit and berries.

Nightingales are summer visitors to Europe, spending the winter in Africa. Being single brooded, they leave from late July, earlier than other migrants. They feed on the ground, mainly on insects and worms, and also berries.

A close relative, the thrush nightingale, *Luscinia luscinia*, is found in eastern Europe; it occurs as a vagrant farther west. It has greyer underparts than the nightingale and its song is even more powerful, although it lacks the rising crescendo.

It is impossible to ignore the song of the nightingale and it is equally difficult to see it. The nightingale, when it is spotted, can look disappointingly drab, but when it catches the light the rich brown plumage and chestnut tail is, in fact, quite attractive. Nightingales habitually sing at night, from mid-April to mid-June, their loud phrases carrying some distance on a warm summer evening, but they do sing in the day as well.

BLUETHROAT *Luscinia svecica*

THIS CLOSE RELATIVE of the nightingale has a similarly beautiful song, but, unlike its cousin, looks far from nondescript. The male's throat colour is the most striking feature – bright blue, bordered with black and red bands on the upper breast, and either a red or white spot in the centre of the blue. The pale buff stripe over the eye and orange outer tail feathers add further distinction. But for all their showy appearance, bluethroats can be difficult to see, spending much time skulking in undergrowth or on the ground. They stand even more upright than a robin and frequently cock and flick their tails to show the orange patches. These birds both run and hop on the ground, and if disturbed they fly fast and low to the nearest hiding place.

Bluethroats breed in marshy scrub, riverside woodland, reedbed edges and also on dry, vegetated hill slopes such as heaths. The nest is built in a hollow or bank in damp ground and from five to seven eggs are laid, with only one brood raised. On migration they favour marshy areas and fields with scrub.

Unlike nightingales, bluethroats sing from a prominent perch such as a bush, tree or telegraph wire. They have a song flight similar to the pipits, rising into the air and floating down with wings and tail spread. The song is varied and musical and repeats phrases, some of which are similar to a nightingale's but weaker, while others are trilling, hissing or bell-like. The song often includes phrases copied from other species like redwing, icterine warbler and reed bunting. Bluethroats sing at night, but not as habitually as the nightingale. Calls include a nightingale-like *hweet*, a sharp *tacc, tacc* and a croaking *turrc, turrc*.

The bluethroat is a summer visitor to Europe and occurs in two distinct forms. The red-spotted race breeds in northern Europe and the white-spotted race breeds in central and parts of southern Europe. Both races winter in North Africa and Arabia. The diet is mainly insects with some snails and worms, and seeds and berries in the autumn.

FACTS AND FEATURES

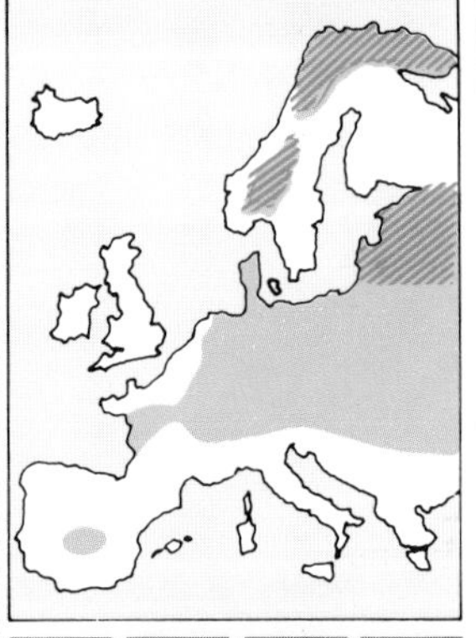

torr-torr-torr-torr, ting-ting-ting

Song A loud, varied and musical song with repeated phrases, some trilling and others bell-like.
Behaviour Sings from a prominent song post on a bush, tree or telephone wire.
Habitat Marshy thickets, scrub and heathland.
Nest Of grass and roots in a hollow or on a marshy bank in swampy scrub. Five to seven eggs, one brood.
Food Flies, beetles, aquatic insects, snails, worms and some seeds.

shows the distribution of the red-spotted bluethroat.
shows the distribution of the white-spotted bluethroat.

Perhaps the most strikingly coloured of their family, the male bluethroats have a bright blue throat with a red or white spot in the centre and a red lower border. Females lack the colour and have brown markings on the throat, but both sexes have a pale stripe over the eye and orange outer tail feathers.

BLACK REDSTART *Phoenicurus ochruros*

The male black redstart has black underparts and a grey back with a distinctive white wing patch, but will start to breed while the plumage is still immature; when young the males resemble the females with grey underparts and a grey-brown back. In both young and mature birds, of both sexes, the tail and rump are reddish-orange. Black redstarts are more likely to be seen than the redstart in both built-up areas and on open, rocky hillsides.

FACTS AND FEATURES

swer-swee-sweeoo

Song A short, far-carrying warbling phrase, followed by a burst of strange clacking, crackling sounds.
Behaviour Sings from a prominent perch at the top of a building, tree or rocky outcrop.
Habitat Buildings, rocky slopes and cliffs.
Nest Of grass and moss in a hole in a rock or building, on a ledge or shed rafter. Four to six eggs, one or two broods.
Food Small beetles, flies, ants, caterpillars, spiders and berries.

THE BLACK REDSTART is a fairly recent addition to the list of British breeding birds, only becoming established during this century, and even now breeding in small numbers. It is similar in size and shape to its commoner cousin, the redstart, and shares with it the orange-red tail from which both species derive their names. Unlike the redstart, the black redstart spends much time on the ground, where it hops or runs when searching for food. It likes perching on rocks and walls, where it stands upright, flicking its tail. It can be shy and is often only glimpsed as a grey-black bird with a reddish tail as it flits away.

In the breeding season, black redstarts are found from sea level to mountains; they frequent rocky slopes, cliffs and gullies. In Britain and north-western Europe they also live in urban areas around large buildings, ruins and construction sites.

The song is always delivered from an elevated song post such as the roof of a building, top of a tree or rocky pinnacle. It is a short and fairly simple warbling phrase with remarkable carrying power, followed by a strange grating, rattling sound rather like ball-bearings being shaken together. Singing begins in early April and can continue to July, or even September. Call notes are a short *tsip* and a *tic-tic* of alarm.

The black redstart's nest is built in a hole in rocks, or on the beam of an outdoor shed or on the ledge of a building. It has not been known to use a nestbox in Britain, but might favour the open-fronted type. Four to six eggs are laid and two broods are often raised.

In Britain, this species is a summer breeding visitor, a passage migrant and a winter visitor. Breeding records are concentrated in central, eastern and south-eastern England. British wintering birds are seen mainly around the south and west coasts of England and Wales, and on the east coast of Ireland. The main foods are insects, spiders and some berries.

REDSTART *Phoenicurus phoenicurus*

The fiery orange-red tail of the redstart is conspicuous in flight. Males are more colourful than females, with a black throat and face, orange breast and grey back and crown with a white forehead.

Females have orange-buff underparts and grey-brown upperparts. A woodland species, the redstart spends less time on the ground than the black redstart.

WITH A FLASH OF ORANGE, a bird flies from a low branch of a woodland oak – this is often the first view of a redstart in its breeding habitat. It favours mature deciduous woodland, but is also found in riverside willows, old orchards, wooded parks and pine forest.

The redstart spends most of its time feeding in the tree canopy, eating mainly insect larvae and spiders. It feeds actively, flitting between branches and flying out after insects in the air. This bird also perches on a branch or fallen log, then flies down to the ground and hops about in a robin-like fashion in search of insects, before returning to the branch. When perched, the redstart frequently quivers its tail, earning the country names of 'fire-flirt' and 'flirt-tail'. Males are more colourful than females, having a black face and throat and orange breast.

The redstart's brief song is composed of short phrases, often reminiscent of a robin, which die away in a jangled jumble of notes. Some songs can be far-carrying while others are barely noticeable; likewise, the singing bird may be hidden in the tree canopy or sitting exposed on a branch. It moves from tree to tree when singing and there is often a considerable interval between bouts of song, making an individual difficult to locate. Singing commences in April and continues into June. The characteristic calls, often heard before the bird is seen, are a clear *whee-tuc-tuc* and a plaintive chiffchaff-like *hweet*. The male has a courtship flight, with tail flashing, and a perched display with drooping wings and tail fanned.

Nesting occurs in holes and other cavities, in a tree or stump at almost any level, or in a stone wall. Old woodpecker nest holes are often used, and hole-fronted nestboxes readily attract redstarts. A garden with a mature orchard and woodland nearby is often suitable for breeding. Six eggs are normally laid and two broods are common.

Redstarts spend winter in Africa and arrive in Britain for the breeding season from early April to early June. The return migration may begin in late July and continues until October. In autumn, the east and south coasts of Britain occasionally witness large 'falls' of redstarts, when birds migrating south are blown across the North Sea in bad weather. The main wintering area is the Sahel region, south of the Sahara, where drought conditions have caused populations to decrease.

FACTS AND FEATURES

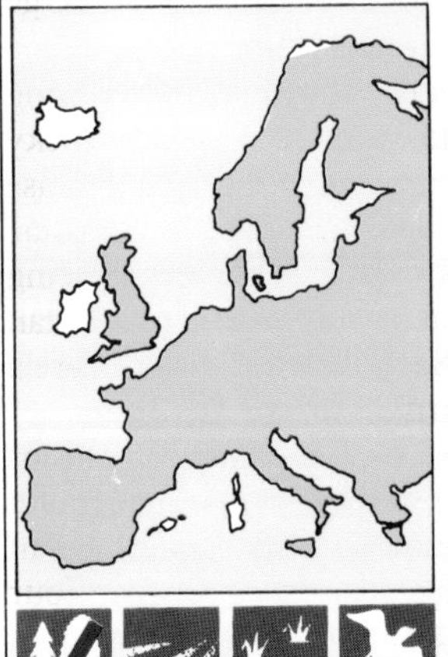

whoo, wheee, too-too-too, tsweee

Song Short robin-like phrases of melodious warbling, finishing with a jangle of notes.
Behaviour Sings from thick tree cover and from an exposed branch, moving position between songs.
Habitat Woodland, parkland and orchards.
Nest Of grass, bark, moss and roots in a tree hole, wall crevice or hole-fronted nestbox. Six eggs, often two broods.
Food Beetles, caterpillars, flies, spiders, worms and some berries.

WHINCHAT *Saxicola rubetra*

The long, broad, pale stripe above the eye distinguishes the whinchat from the stonechat. When perched and in flight it also shows white sides to the base of the short tail. Its underparts are orange-buff and its back is streaked with pale and dark brown. The males have dark sides to the face; females are paler brown.

FACTS AND FEATURES

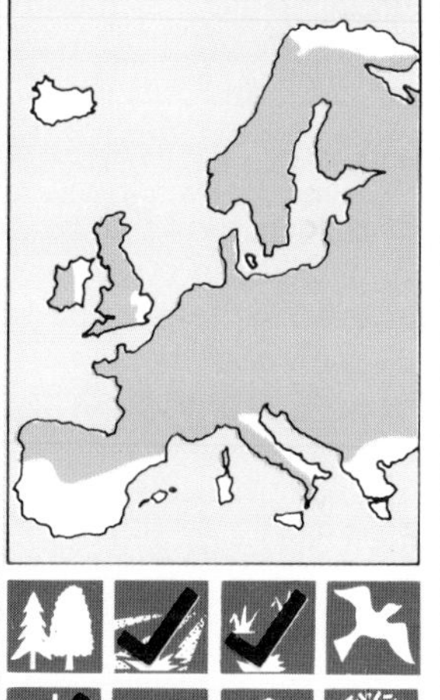

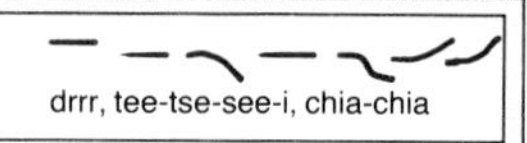

Song A brief, variable warble with pauses between each phrase, *tee-tse... tee-tse-see-i... tee-tse-see-i*.
Behaviour Sings prominently from top of bush or tree, on the ground or rarely in display flight.
Habitat Open country, meadows and commons with bushes.
Nest Of grass and moss, built on the ground among tall grass or under a bush. Five or six eggs, two broods.
Food Beetles, flies, caterpillars, spiders and worms, taken mostly from plants.

THE DIMINUTIVE WHINCHAT, and its cousin the stonechat, are the smallest members of the thrush family in Europe. The whinchat breeds on open grassland, bracken-covered heaths and young forestry plantations, where it always perches prominently on a bush, fence or piece of scrub. The male has a dark head broken by a long, broad, white stripe stretching from above the bill and over the eye almost to the back of the neck. In flight there is a white wing mark and white sides (absent in the stonechat) to the short tail. On the ground, these birds stand very upright and hop along, and when perched on a bush they flick their wings and bob. They are insectivorous, taking their prey from flowers and grasses and occasionally in flight. Spiders, worms and small molluscs may be eaten in the autumn.

The male perches on the top of a bush or tree to sing, although sometimes he sings from the ground, and rarely during a tree pipit-like aerial display. The song is a sweet and rather mechanical warble, sounding at times similar to a robin, redstart or wheatear. It is quite far-carrying and consists of short phrases with an interval between each *tee-tse... tee-tse-see-i... tee-tse-see-i*. The male also sings while displaying to the female with his wings quivering, tail fanned and head held back. He has a short song period from late April to the end of June. The chief call is a scolding *tic-tic* or *tu-tic-tic*, as well as churring and rattling notes.

Whinchats return to Europe for the summer from late April to late May. Both sexes arrive at about the same time and territories are rapidly established. The species is ground-nesting, building among tall grass or at the foot of a bush. Five or six eggs are laid and two broods are often reared. The adults are protective towards their nest and perch anxiously on nearby bushes, giving their *tic-tic* call if danger threatens.

The birds begin the journey to their wintering grounds, in tropical Africa, during August; many will be south of the Sahara by early October. Scandinavian breeding birds are commonly seen along the east coast of Britain in May, late August and September, especially after easterly winds. On migration they are seen in reedbeds, saltmarshes and on open shores.

STONECHAT *Saxicola torquata*

FACTS AND FEATURES

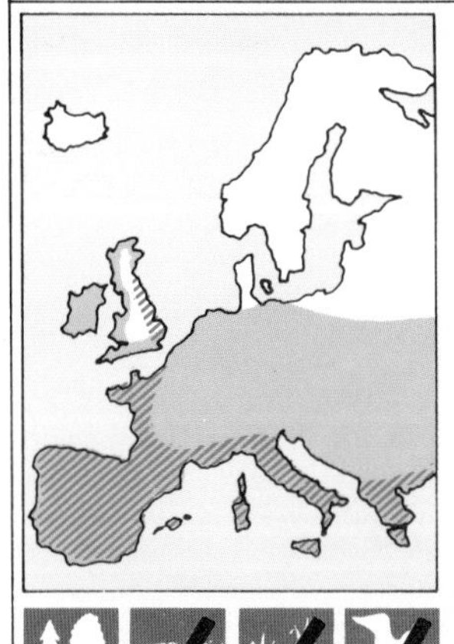

weetsoovi, tsweeoo-tsweeoo

Song A variable dunnock-like warbling of a series of double notes, rising and falling.
Behaviour Sings from a perch on top of a bush or in a dancing vertical song flight.
Habitat Commons, heaths and coastal bushes.
Nest Of moss, grass, gorse and wool on the ground at the base of a bush. Five or six eggs, two or three broods.
Food Beetles, caterpillars, flies, grubs, spiders and worms taken on the ground.

IN CONTINENTAL EUROPE, stonechats are found mainly on scrub-covered hillsides, vineyards and meadows. In Britain, they are more confined to coastal areas with gorse, heather, grass and bracken. Wherever they live, they are difficult to ignore, since they perch high on gorse or other bushes and are extremely vociferous. The male is unmistakable with his black head, white half-collar and chestnut underparts. The lack of a stripe above the eye distinguishes the stonechat from its relative, the whinchat. It perches upright and hops on the ground in a robin-like way.

The song of the stonechat is short and warbling and resembles that of a dunnock. Each song is a series of double notes, the first clear and sharp and the second deeper, with the whole series rising and falling. It is delivered either from a bush or in a song flight. The male flies from his perch to a height of 50-100 feet (15-30 metres) and 'dances' up and down while singing, each phrase uttered as he drops a couple of feet. He sings from mid-February until the end of June. The call note gives the stonechat its name, a hard *tack, tack* that sounds just like two stones being knocked together. The bird also gives a plaintive *hooeet*, which it may join onto the other call.

The nest is built on the ground at the base of a thick clump of gorse or a similar bush, sometimes with an entrance tunnel through surrounding grass. Five or six eggs are laid.

The all-black head of the male stonechat contrasts with the white sides to its neck; its underparts are chestnut-brown. Females are duller-looking, with streaky brown upperparts. The lack of the pale eyestripe and its dark throat distinguishes it clearly from the whinchat. The stonechat perches upright and has a noisy tacking call.

Stonechats are absent from Scandinavia and much of northern Europe. Central European birds are migratory, spending the winter along the Atlantic coast of Spain or around the Mediterranean. British stonechats are more sedentary, staying close to their breeding territories – although some northern birds move closer to the coast in winter. They feed largely on insects and spiders, mainly taken on the ground. This diet, and the species' residential habits, mean that it suffers badly in severe winters when snow covers the ground. However, after a very hard winter it is able to build up its numbers rapidly by raising three broods instead of the more usual two.

WHEATEAR *Oenanthe oenanthe*

THE WHEATEAR'S MISLEADING NAME comes from the Anglo-Saxon words for 'white rump' and has nothing to do with grain. Sometimes called the common or European wheatear, it is a typical chat, perching very upright, but preferring boulders to bushes. This active bird hops rapidly along the ground, sometimes pausing on a rock and bobbing or cocking its tail. The males have an attractive black mask, black wings, a grey crown and back, and buff underparts. In flight the white rump contrasts markedly with the black tail. Wheatears breed on upland pastures, moorland, mountains, downland, shingle, dunes and rocky areas.

The song is a loud warble which contains harsh chatters delivered in short phrases. It sometimes suggests a whitethroat or a lark, and is more forceful than the songs of its relatives. Song perches are usually stones, lumps of earth, walls and fences; occasionally the bird adopts a song flight with fluttering wings and fanned tail. Wheatears sing from April to June and are sometimes heard at night. They have a loud call, a hard *chack, chack* sometimes preceded by a high *wheet*.

Being a ground-dweller, the wheatear's nest is also on the ground. Cracks in rocks, holes beneath boulders and rabbit holes are favourite sites, and it also uses artificial sites in old pipes, cans and kettles. There are usually six eggs and one or two broods.

The North African race of wheatear is distinct from the European forms, having a black throat. European wheatears return from Africa in late February and March. They are one of the first migrants to arrive, and also one of the greatest travellers. Some pass through Europe on their way to Greenland and Canada. These are of a larger and paler race than the truly European birds and tend to perch more on bushes when on migration. Some of them 'island hop' through the British Isles, the Faeroes and Iceland before reaching Greenland, while others fly the 2,000 miles (nearly 3,000 kilometres) from Spain to Greenland in one journey. They feed on a diet of insects.

FACTS AND FEATURES

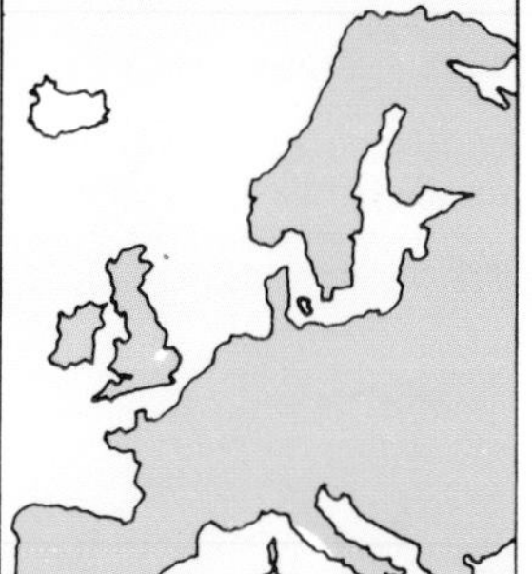

Song A loud warble with lark-like and harsh chattering notes.
Behaviour Sings from a rock, wall or fence, in normal flight and in a song flight.
Habitat Grassland, moorland and dunes.
Nest Mainly of grass, hidden in a hole under a rock, in a rabbit burrow or in a specially buried box. Six eggs, one or two broods.
Food Beetles, flies, bees, ants, grasshoppers, spiders and small snails.

The bold black-and-white tail and rump pattern of the wheatear is the most striking feature of both this and other wheatears. The wheatear has a white rump and white edges to the base of the black tail, forming a 'T' shape, which is particularly distinctive in flight. Males have a blue-grey back and crown, a handsome black face mask and orange-buff underparts.

BLACK-EARED WHEATEAR *Oenanthe hispanica*

A Mediterranean species, the black-eared wheatear shows more white in the tail than the wheatear. Males can have sandy-buff or almost white upperparts and they may also have a black face patch with a pale throat, or a black throat and face. The wings appear very black. Females are similar to female wheatears but have darker wings and lack the pale stripe over the eye.

THE HANDSOME black-eared wheatear is a summer visitor to southern Europe and exists in a number of different forms. Males can appear almost white in the east, while western birds are a more golden-rufous colour. All forms have a black face mask, but some also have a black throat. Like all wheatears they display a white rump, which extends more onto the tail than in the common (European) wheatear.

The black-eared wheatear is found in the dry, stony lowlands and heathlands of the south, while its cousin the wheatear is confined to rocky mountain areas. The black-eared species tends to perch more on wires and bushes than the wheatear, but otherwise behaves similarly, bobbing up and down when alarmed and uttering its call note.

The song is similar to that of the wheatear, with high-pitched phrases, *schwer, schwee, schwee-oo*. Favourite song perches are rocks and telegraph wires, and there is sometimes a brief, circling song flight with jerky undulations. The call note is unlike a wheatear's, being a low, grating *gsch*, sometimes followed by a plaintive, whistling *kscheup*, and a rasping *schrrr*.

The nest is built on a hillside, in a depression under a rock or in a recess overhung by gorse. Piles of stones and old ruins are also used, and artificial sites such as an old bucket are not unknown. Four or five eggs are laid and in some parts of Europe two broods are raised.

Black-eared wheatears spend the winter in Africa south of the Sahara. They are often seen on passage in large numbers, on Mediterranean islands such as Mallorca. Black-eared wheatears are occasionally seen as rare vagrants farther north, including Britain. They feed mainly on insects picked from the ground.

FACTS AND FEATURES

schwer, schwee, schwee-oo

Song Similar to a wheatear's with more high-pitched phrases including a distinctive *schwer, schwee, schwee-oo*.
Behaviour Perches on wires to sing, also on rocks and has a circling song flight.
Habitat Dry, stony country and rocky hillsides with trees.
Nest Of grass and moss in a hollow under a rock or overhanging bush. Four or five eggs, two broods.
Food Beetles, grasshoppers, flies and ants.

ROCK THRUSH *Monticola saxatilis*

THE MALE of this medium-sized, rather plump, short-tailed thrush is strikingly coloured with a blue head and back, and orange underparts and tail – when flying away, the flash of tail colour makes it look like a large redstart. This species spends a great deal of time on the ground, moving by hopping, and perches on rocks, buildings and trees, standing upright like many other thrushes and chats. It is a shy and solitary bird, and if disturbed it may jerk its tail up and then lower it with a swinging motion before disappearing behind a boulder or rock face. It breeds in high rocky areas with some scattered trees, but may occasionally be found on lower hills and around old buildings. In winter, in Africa, it also frequents stony areas and sometimes cultivated fields or the open savanna.

The mellow and warbling song is delivered from a rock or telegraph post. There is also a song flight in which the male flies up with slow wing beats to a considerable height, then fans his tail and spreads his wings as he circles downwards to his perch while singing. The call is a hard *chack, chack*.

The rock thrush's nest is built in a hole among rocks or in a wall, or sometimes in a bank or tree hole. Four or five eggs are laid and a single brood is raised.

Unlike its larger relative the blue rock thrush, the rock thrush is a summer visitor to southern Europe. Birds arrive in April and May, males earlier than females. Departure in the autumn takes place in August and September; western European populations go to West Africa, and eastern European and Asian birds go to tropical East Africa. They feed on insects, especially beetles, crickets, flies and butterflies, watching for prey from a perch on a rock or digging it from the ground. Snails, worms and even frogs and lizards have been recorded as food items.

FACTS AND FEATURES

dlu-dlu-dlu, dooi, dooi,
chivichoo...

Song A loud, fluty warbling, sometimes imitating other species.
Behaviour Sings from a rock or telegraph post and also in a circling song flight.
Habitat Open rocky areas with trees.
Nest Of grass, roots and moss in a hole in a rock, wall, bank or tree. Four or five eggs, one brood.
Food Beetles, grasshoppers, flies, moths, butterflies, spiders, snails, worms and berries.

The male rock thrush is unmistakable, with orange underparts and tail, and a blue throat, head, back and rump. The back carries a bold white patch. Females are mottled with pale-edged feathers, and resemble a large speckled female redstart. Both sexes are shy and solitary birds.

BLUE ROCK THRUSH *Monticola solitarius*

In good light the blue body colour of the male blue rock thrush can be striking, but it often appears a blackish-blue. In song, the blue rock thrush will usually be found perched on the top of a high rock, from which it chases after insect food. Females are a dull brown colour, with mottled underparts. The blue rock thrush is secretive and takes flight when approached.

THE BLUE ROCK THRUSH is found in southern Europe, from mountainous regions down to sea level. It is a shy and solitary bird, and when disturbed it bows and flits its tail before escaping with a darting flight behind a rock. In bright light the blue body of the male contrasts with its slaty-black wings and tail, although in duller conditions it can look much darker and resemble a small blackbird. When flying across a valley the blue rock thrush has a direct thrush-like flight with pointed wings, and when perched it stands upright.

From a vantage point on a rock pinnacle or the top of a wall, this thrush watches for food, either pouncing onto the ground or flying up after a passing insect. The species has a varied diet consisting of insects (mainly taken on the ground), spiders, worms and snails. It sometimes catches small mice and lizards, and coastal birds feed in tidal areas on marine worms, sea snails and small crustaceans.

The song is very thrush-like, similar to a blackbird or mistle thrush, containing fluty phrases such as *choo-sri, churr-titi* repeated with long pauses in between. Song is usually given from a perch, such as on top of a rock, a ledge on a rock face, the branch of a tree or the top of an electricity pylon. The bird also has a vertical song flight and can be heard singing while flying across a valley. Its call notes, usually uttered when disturbed or alarmed, include a chat-like *tac-tac* or *tchuck*, a high *tsii* and a nuthatch-like *hiut-huit*.

The nest is sited in a dark rock crevice, a hole in a wall or cliff, under stones or in a cave. Five eggs are normally laid and one brood, sometimes two, is reared.

Blue rock thrushes are resident in southern Europe, in some areas moving to lower ground in winter, when they may be found in old quarries, ruined buildings and on rock piles.

FACTS AND FEATURES

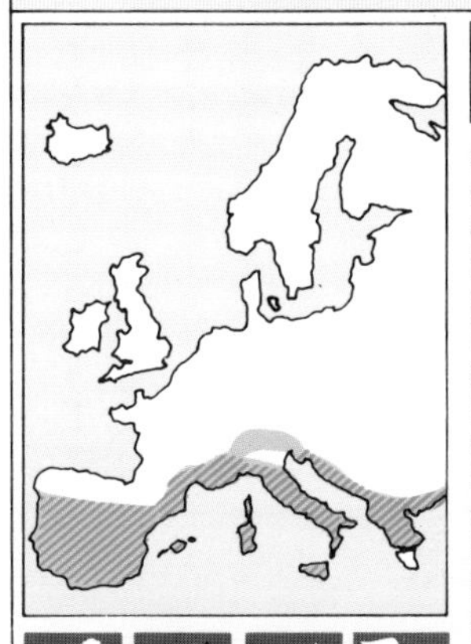

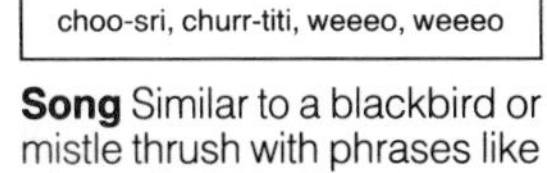

Song Similar to a blackbird or mistle thrush with phrases like *choo-sri, churr-titi*.
Behaviour Sings prominently from a rock, ledge, branch or pylon and in a vertical song flight.
Habitat Rocky hillsides and cliffs.
Nest Of grass and roots in a rock hole, wall, cliff or cave. Five eggs, one or two broods.
Food Caterpillars, grasshoppers, spiders, worms and snails. Occasionally small mice and lizards.

BLACKBIRD *Turdus merula*

A common garden bird, the all-black colour of the male is broken only by its yellow-orange bill and eye-ring. Its song has the typical fluty warbling of the thrush family and it is one of the first birds in the dawn chorus. Young males are browner, lack the eye ring and have dark bills.

THE BLACKBIRD IS THE TRUE 'early bird', being first to sing each day in most areas. Its familiar clear, fluty warbling is heard from garden to woodland to upland scrub. It is one of Europe's commonest birds and a typical thrush, with its long tail making it perch less upright than its cousins, the chats. The male's glossy black plumage and yellow-orange bill are instantly recognizable on a roof top or in a field.

On the ground the blackbird both runs and hops, pausing and sometimes raising its tail. When landing, its tail is cocked and wings drooped slightly; when alarmed, it calls and flicks its tail and wings. The blackbird usually feeds not far from cover, ready to dash under a bush if disturbed.

The song is a fluty warble with mellow phrases and a clear quality. Certain phrases may be used repeatedly, but the repertoire is generally varied. Typically, a blackbird sings from a tree or other prominent perch such as a roof or TV aerial. It can be heard throughout the day, but is particularly vocal at dawn and dusk, and occasionally at night. These birds sing from February to July, with a quiet sub-song in autumn and late winter. There are a number of calls: a low *chook, chook, chook* when disturbed, but a loud ringing chatter with a hysterical quality when alarmed. If mobbing an owl, cat or other predator, the birds give a scolding and persistent *chik-chik-chik*. Another call, when perched and in flight, is a thin *tsee*.

Blackbirds nest in bushes, hedges, trees and sometimes on the ground. They do not enter nest-boxes, but often use shelves in sheds. They lay from three to five eggs and have up to three broods, although in mild winters the early breeders may have more.

Blackbirds are resident across Europe, with the exception of Scandinavia. There they are summer visitors, migrating south in winter; many come to Britain, arriving in October and November and returning north in March and April. The varied diet consists of fruit, seeds, worms and insects.

FACTS AND FEATURES

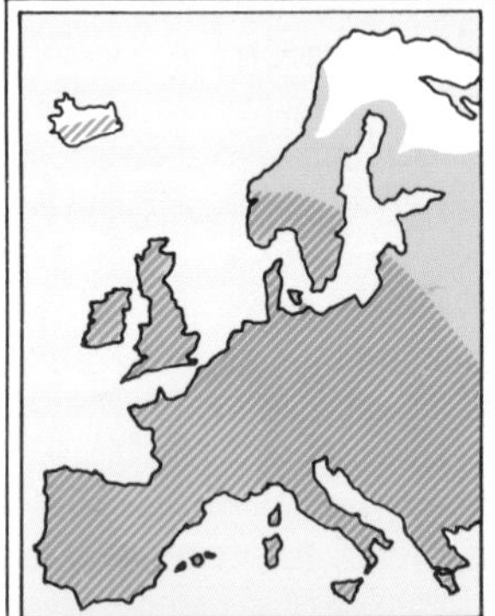

choo-veeoo, weeoo-choo, choo-titi

Song A mellow, fluty warbling with varied phrases. Imitates other sounds.
Behaviour Sings from a tree, roof or TV aerial, sometimes at night.
Habitat Gardens, woodland and hedges.
Nest Of grass, moss and mud in a hedge, bush or creeper. May nest on ledges in buildings. Three to five eggs, two or three broods.
Food Worms, caterpillars, ants, spiders and snails. Apples, strawberries, raspberries and wild berries. Readily visits a bird table for a variety of foods.

Ivy growing in the garden can attract many species of birds. This blackbird (left) *is gorging himself on the berries. The species will also nest in ivy, as will many others and, although regarded as an unsuitable garden plant, it rarely does any damage. Blackbirds are also fond of the berries of cotoneaster, barbary, honeysuckle and hawthorn.*

Sheds are a favourite place for blackbirds to build their nests in (above) *but, if they take up residence, you must be sure that you do not accidentally block up their entry point or leave the door open to let predators like cats in.*

The clutch is usually four or five but larger numbers of eggs have been recorded. The inner layer of mud has been covered by a lining of dry grass (right). *These eggs are typically marked, but coloration is variable.*

FIELDFARE *Turdus pilaris*

A winter visitor to much of Europe, where it can be seen in flocks, feeding in open fields, the fieldfare is easily recognized by its grey head and rump, chestnut back and black tail, black speckled breast and flanks and white belly. Fieldfares will plunder gardens for food in severe winters, eating berries and apples.

FIELDFARES ARE LARGE, bold thrushes with a grey head, chestnut back, grey rump, black tail and black spotted breast. They are gregarious in the breeding season, nesting in colonies – preferably in small woods of pine and birch with fields and clearings nearby, but also in large forests, and even on the ground if there are no trees. In orchards they can become a pest, taking the fruit, and in parts of Europe they have become a garden bird rather like the blackbird.

In Britain, the sight of a loose flock of fieldfares flying over, giving their characteristic 'chacking' call, is a familiar winter sight. British individuals come from Scandinavia and their numbers depend on the availability of food there. In a good year, when berries are plentiful and there is little competition from birds farther east, they may not arrive in Britain in numbers until well into winter. With a food shortage, they may start arriving in September.

The fieldfare's upright stance and distinctive colouring make it easily recognized on the ground, while in flight the call and the flash of white under the wings is enough to identify this species. They are aggressive birds, both on their breeding and winter territories.

The song is not very musical, unlike most other thrush songs, and consists of chattering and squeaking with phrases like *took-took-cherri-weeoo*. It can sometimes be heard on fine spring days in Britain, before the birds have migrated north. The call is a loud chattering *chack-chack-chack* in flight and when perched; if disturbed, the bird may also give a quiet *seee*.

Fieldfares build their nests in tree forks, sometimes only a few feet from the ground. They lay between four and seven eggs. In the south two broods are normal, while in the north there is only time for one.

In Europe, fieldfares are resident in the east, summer breeding visitors in the north, and winter visitors in the west and south. Migrating birds move both south and west and this has helped them to colonize as far as Iceland and Greenland. They feed on worms, insects, slugs and berries. In severe winters when snow covers the ground they readily visit gardens to seek out apples and berries.

FACTS AND FEATURES

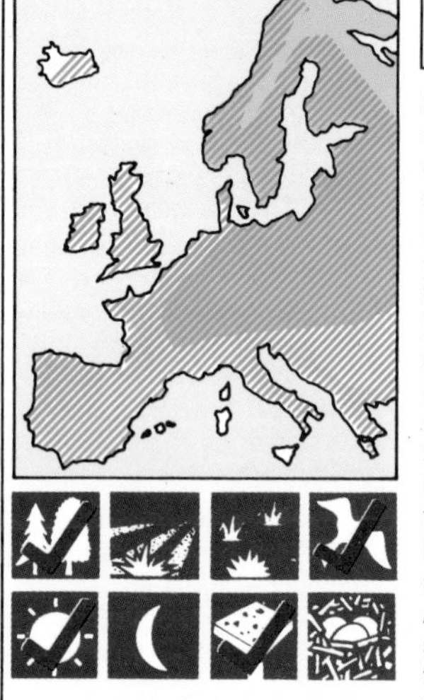

took-took-cherri-weeoo

Song An unmusical chattering warble, *took-took-cherri-weeoo*, with squeaks and chacking noises.
Behaviour Sings from a perch in a tree and when in flight.
Habitat Woodland, orchards and open country.
Nest Of grass, twigs and mud in the fork of a tree, sometimes against the trunk. Four to seven eggs, one or two broods.
Food Beetles, flies, spiders, snails, slugs and worms. Apples, wild berries and some seeds. Visits gardens for fruits and berries in hard winters.

SONG THRUSH *Turdus philomelos*

A common resident of gardens and woodland, the song thrush has warm-brown upperparts and a black-speckled buff breast. Song thrushes can sometimes be located by the loud tapping sound they make when trying to break open snail shells on rocks or stones. Singing birds perch high on trees or TV aerials, repeating each of their song phrases two or three times. Male and female look identical.

FACTS AND FEATURES

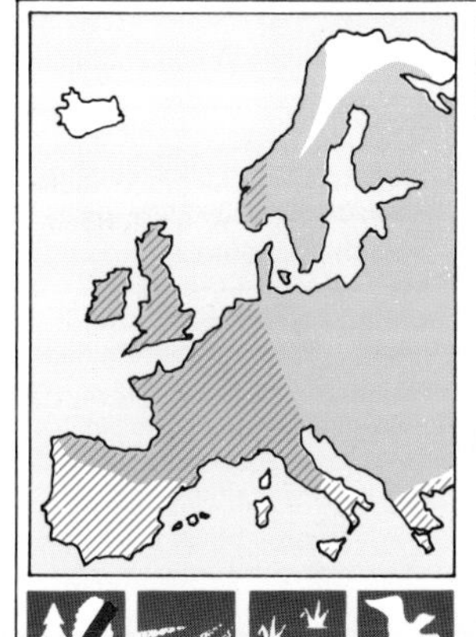

...cherwee, cherwee, cherwee...

Song A loud series of short musical phrases, each repeated a few times.
Behaviour Sings from a song perch on a tall tree, building or bush.
Habitat Gardens, woodland and hedges.
Nest Of grass, roots and twigs in hedges, bushes and trees. Four or five eggs, two or three broods.
Food Worms and snails, also beetles, flies, caterpillars, spiders, fruit and berries.

WITH ITS BROWN BACK and black-speckled breast, the song thrush is recognized by many people. Its song is also widely known and often acts as an early morning alarm. A common inhabitant of gardens, parks and woodland, the song thrush likes open grassy areas. It has an upright stance and runs or hops along the ground, often holding its head on one side to listen and look when searching for food.

Perched high in a tree or on a TV aerial, the loud and cheerful song begins at first light. Its structure makes it easily recognizable: simple phrases are repeated two or three times, although with this bird's large repertoire of such phrases, the song can be extremely varied. Song thrushes are good mimics, and as well as copying the calls of redshank, ringed plover and green woodpecker, they have learnt to imitate man-made sounds like the trill of a modern telephone. They start to sing in earnest in January, after males have established territories in late autumn and early winter. On cold days they may go quiet, but singing occurs on most days by March and reaches a peak in April and May, becoming silent by the end of July and starting again in October. Two main call notes can be heard, a short *tick* and a softer, longer *sipp*.

Nests are built in trees, hedges and bushes, or sometimes in a shed or a hole in a wall. Thick foliage is preferred – holly, elder, hawthorn, conifers and creepers are favourites. In areas with little tree cover, nests may be situated in heather or bracken. Four or five eggs are commonly laid with up to three broods.

Song thrushes are resident in Britain. However, birds from the Scottish highlands and some upland areas of England and Wales migrate in winter to Ireland, France and Spain. Birds from Belgium and Holland winter in southern England, while northern European birds fly as far as southern France and Spain. For most of the year they eat worms and snails and are famous for their technique of using a stone or other hard surface as an 'anvil' for smashing snail shells. In summer they eat some insects, and in autumn and early winter berries are an important food.

REDWING *Turdus iliacus*

Often seen in large winter flocks in the company of fieldfares, feeding in open fields and along hedgerows, redwings resemble small song thrushes but have bright orange-red flanks and distinctive pale eye-stripes and have streaked rather than speckled breasts. They are nocturnal migrants and their thin, loud seeip *call can often be heard on a clear autumn night when they arrive in Britain.*

A SMALLER RELATIVE of the song thrush, the redwing is a winter visitor to all but northern Europe and is the only true thrush to breed in Iceland. At a glance it could be mistaken for a song thrush, but a closer look reveals its striking face pattern with a bold stripe above the eye and its streaked, rather than spotted, breast. The red flanks show below its closed wing, and in flight the full extent of its red underwing is revealed. Redwings are gregarious in winter and often gather in mixed flocks with their larger cousins, the fieldfares.

The song of the redwing is infrequently heard in Britain, where it rarely breeds. It is a simple song of repeated phrases with clear, fluty notes, *trui-trui-trui* or *teetra-teetra-teetra*, and a chuckling warble at the end. It also has a sub-song, often given by groups as they gather in spring before migrating north. This consists of a twittering warble, sometimes incorporating some fluty notes reminiscent of the main song. The call is a thin, loud *seeeip*, often given during flight. On clear October nights when redwings are arriving in Britain, the sound of their calls as they fly over is a reminder of the winter to come. If disturbed in a hedge they have a harsh rattling call.

Redwings nest in mixed woodland, gardens and parks, low down in birch trees, on tree stumps, in evergreen shrubs and sometimes on the ground, and even in walls and old sheds. They lay five or six eggs, usually in June and July, and raise two broods.

The bulk of Europe's redwings breed in Scandinavia and Iceland, with a few pairs in Scotland. They migrate south in the winter, but have no set pattern, their destinations often dictated by weather conditions and food supplies. In general, Scandinavian birds come to England and continental Europe, while Icelandic birds visit Scotland and Ireland. In winter they rely on worms and fruit for food, and flocks can be seen searching fields together or stripping hawthorn bushes of their berries. Sudden snow or lack of food causes them to move on until, in March and April, they prepare to return north. In very cold conditions these birds readily visit gardens for cotoneaster or pyracantha berries, and apples also attract them.

FACTS AND FEATURES

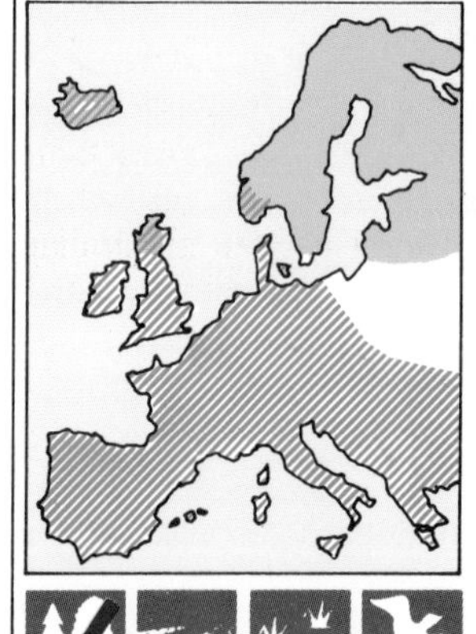

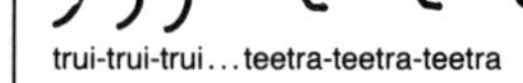

Song A phrase of three or four fluty notes, *trui-trui-trui*.
Behaviour Sings from trees, bushes and scrub.
Habitat Birch woodland, bushes and open ground.
Nest Of grass, twigs and earth in a tree, bush or sometimes on the ground. Five to six eggs, two broods.
Food Worms, snails, beetles, flies, caterpillars, fruit and berries.

MISTLE THRUSH *Turdus viscivorus*

The mistle thrush is a large thrush differing from the song thrush in its greyer upperparts, large bold spots on the breast, paler rump and white-tipped tail. The flight pattern is distinctive with slower wingbeats, making it more undulating and showing the bird's white underwings. It has a loud rattling call and will aggressively defend its food sources in winter.

A LARGE, BOLDLY SPOTTED bird, greyer than its close cousin the song thrush, the mistle thrush has gained its name from its supposed liking for mistletoe berries. It stands very upright and is more assertive and aggressive than its smaller relative. In flight it closes its wings markedly, which makes it undulate rather than fly straight and direct; also in flight, its white underwing shows, distinguishing it from all similar birds except the fieldfare. A close view may show the paler rump and whitish corners to its tail. This species inhabits woodland, parks, gardens and open country. Highly territorial in the breeding season, it often forms flocks in the autumn, but in winter individuals usually defend feeding areas from any other bird.

The song is similar to a blackbird but not as rich. The mistle thrush gives short fluty phrases which show little variation, such as *tee-aw-tee-aw-tee*, but which carry for some distance. Its habit of singing from the top of a tree during a cold, wet and windy day has earned it the nickname 'stormcock'. In fact the mistle thrush begins singing earlier after summer than any of our other thrushes, starting to sing regularly in November. Its peak song occurs in January and February, and by late May, the peak time for blackbirds and song thrushes, it has ceased. The call of the mistle thrush is a distinctive rattling chatter which is given in flight and when perched, being loudest when the bird is excited.

The mistle thrush's nest is usually built in the fork of an oak or other large tree, sometimes up to 33 feet (10 metres) from the ground, but occasionally lower, in bushes like holly and elder. The female normally lays four eggs and usually has two broods, sometimes using the same nest twice in a season.

Northern European birds are migratory, spending the winter in France and Spain, and a few may arrive in Britain in the autumn. Scottish birds move south to Ireland and France, while English birds may travel only a short distance. The diet of the mistle thrush consists mainly of worms, snails and berries, the latter being important in the autumn and early winter. It also eats apple, cherry, plum and pear fruits.

FACTS AND FEATURES

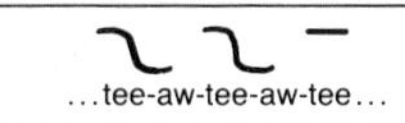

Song Short phrases of fluty, blackbird-like notes, repeated many times, *tee-aw-tee-aw-tee*
Behaviour Sings from upper branches and tops of trees, occasionally in flight.
Habitat Gardens, woodlands, orchards and fields.
Nest Of grass, roots, moss and earth in the fork of a tree at some height. Four eggs, two broods.
Food Fruit, especially cherries, apples and plums, berries like hawthorn, yew and mistletoe. Some snails, worms and insects.

CETTI'S WARBLER *Cettia cetti*

FACTS AND FEATURES

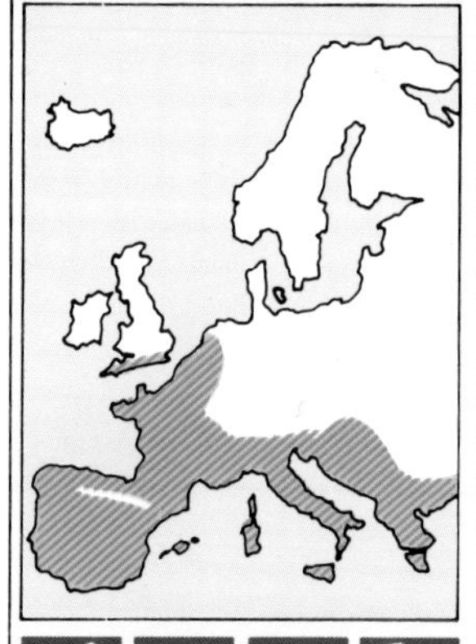

chee, che-weechoo-weechoo-weechoo-wee

Song A loud outburst of *chee, che-weechoo-weechoo-weechoo-wee.*
Behaviour Sings from dense cover of bush or shrub, sometimes at night.
Habitat Bushes and shrubs at edges of rivers, marshes and reedbeds.
Nest Of leaves and grass in a bush or shrub. Four or five eggs, one or occasionally two broods.
Food Flies, beetles, moths, spiders and worms.

THIS SHY and often skulking warbler is a recent addition to Britain's birds and is more likely to be heard than seen. It inhabits bushes, hedges, brambles and scrub along rivers, reedbeds and marshes. It is a small chestnut-brown bird which could be mistaken for a nightingale. The underparts are greyish-white and it has a grey supercilium ('eyebrow'), while its short wings and longish rounded tail give it a distinctive profile. Male and female are identical in plumage, but, unlike other warblers, males have noticeably longer wings and tails than females. This warbler spends most of its time hiding in tangled vegetation, and when glimpsed it is usually a brief view of a brown bird flicking or cocking its tail. It can move quickly and quietly through vegetation and even runs along the ground like a mouse.

The song of the Cetti's warbler is loud and distinctive and is often the first clue to its presence. The rapid burst of notes can be written *chee, che-weechoo-weechoo-weechoo-wee* and are usually delivered from low in a bush. Song can be heard from early in the year to July, when breeding has ceased; nocturnal singing is not uncommon. The call note is a sharp *twiick* or *chick* that may merge into a trill.

Nests are built in hedges, brambles and shrubs, about two feet (less than one metre) from the ground. Males may have up to three partners and each female can lay four or five eggs and have second broods.

The Cetti's warbler has greatly extended its range this century. Originally from the Mediterranean region, it began spreading northwards in the 1920s. By the 1960s it had reached northern France and in 1973 breeding was officially recorded in Britain. Since then numbers have increased, with as many as 300 pairs breeding in Britain in the 1980s. As a resident it is vulnerable to cold winters, and numbers dropped in some parts of Britain after the severe weather in 1981-82. Like all warblers, it feeds on flies, spiders, moths, beetles, caterpillars, snails and worms.

This is a small, shy, wren-like, chestnut-brown bird with a surprisingly loud song. When it is singing from dense undergrowth it can be difficult to spot and is often glimpsed only as it flies into some nearby cover. The rounded tail is longer than a wren's and is reddish-brown above with pale grey underparts. Males and females have identical plumage, but the females have shorter wings and tails.

Fan-Tailed Warbler *Cisticola juncidis*

When it sings its monotonous rasping song this tiny short-tailed warbler is impossible to ignore. In flight it looks as if it is attached to a jerking string. The tail is white tipped and the back and head are streaked with pale and dark stripes. When not singing the fan-tailed warbler keeps to grasses, rushes and other low vegetation.

EUROPE'S SMALLEST WARBLER, the fan-tailed warbler also bears the name 'zitting cisticola', since it belongs to a genus (*Cisticola*) of confusing African grass warblers with names like rattling, piping, whistling, trilling and chirping cisticolas – all referring to their songs. It is a tiny, dumpy, short-tailed bird streaked with rufous-brown, usually seen flitting out of one clump of vegetation and disappearing into another nearby. A close view might reveal the tail feathers clearly tipped with white and the short, rounded wings. When perched it usually holds its tail cocked. Fan-tailed warblers prefer fields and rough grass at the edge of marshes and rivers, but they can be found in drier agricultural areas. Food consists mainly of insects, grubs and caterpillars.

This species, which is likely to be heard before it is seen, has a high rasping *zip, zip, zip, zip*... song which soon becomes monotonous. This may be given when perched, but is more commonly uttered in a bobbing song flight. The call notes are a loud *chip* and an explosive *tew* of alarm.

The nest is a fascinating construction of spiders' webs woven around the stems and leaves of grasses and rushes. It is often long and narrow, sometimes to such an extent that, after entering the nest, the adult cannot turn and has to leave tail-first! The male builds the outside of the nest and may construct several before one is lined and used by the female. Like the Cetti's warbler, male fan-tailed warblers frequently have many mates. Between four and six eggs are laid and two broods, sometimes three, are reared.

FACTS AND FEATURES

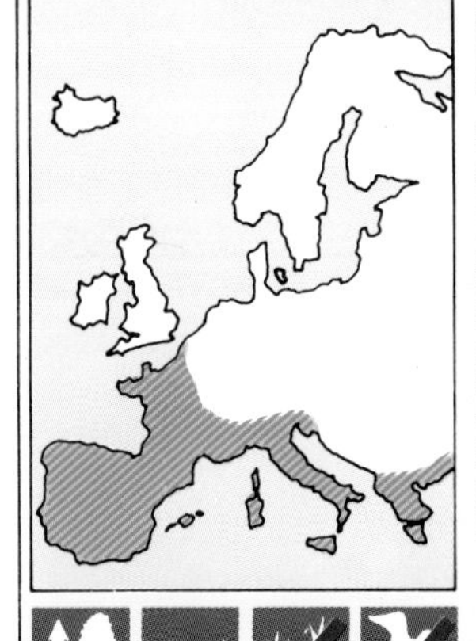

zip... zip... zip... zip...

Song A high, rasping *zip, zip, zip, zip*...
Behaviour Sings from a perch and also in jerky song flight.
Habitat Fields and grass by marshes, lakes and canals.
Nest Of grass and spiders' webs in rushes or long grass. Four to six eggs, two or three broods.
Food Insects and their grubs.

Fan-tailed warblers are resident and mainly sedentary, with any long-distance movements almost certainly undertaken by young birds. They have extended their range in a similar way to Cetti's warblers – expansion across France in the 1970s was spectacular, and in 1976-77 there were two sightings in Britain. Recent winters have forced them back southward, even reducing numbers in the south of France.

GRASSHOPPER WARBLER *Locustella naevia*

BOTH THE ENGLISH and scientific names of this species are derived from its insect-like song, and anyone not familiar with this bird might well believe that they are hearing a grasshopper. A retiring nature makes it difficult to see, but with quiet patience it is usually possible to find this small, slim, olive-brown warbler, with its softly streaked back, sitting in the middle of a bush or reedbed, singing. Its underparts are buffish-grey and can show a yellowish tinge in bright light. When disturbed, it creeps away through the undergrowth and even runs along the ground. Grasshopper warblers are found in bushes and tangled vegetation by wet meadows and marshes, but they can equally well be seen in hedges and bushes in open grassland, and also in young conifer plantations with layers of bramble.

The song is a high reeling trill and has a curious ventriloquial quality – it is difficult to judge the distance of a singing bird, and when it turns its head the volume of sound changes. These warblers are often concealed when singing, but may sit at the top of a bush, when the gaping bill and pulsating throat feathers can be seen as they sing. The trills last from a few seconds to several minutes and carry as far as half a mile (almost one kilometre). Birds sing at night as well as during the day and also on migration, before they reach their breeding site. In England, song begins when they arrive in April and continues until July. The call note is a hard *tchick*.

The nest is built on or close to the ground in tall grass or a bush, and the birds usually creep to it, often through an approach tunnel. Six eggs are normal and birds in southern England may raise two broods, those in the north managing only one. In southern Europe three broods can be reared.

Grasshopper warblers are summer visitors to Europe. They spend the winter in Africa, migrating in August and September. They feed on flies, aphids, moths, caterpillars, mayflies, beetles and spiders.

FACTS AND FEATURES

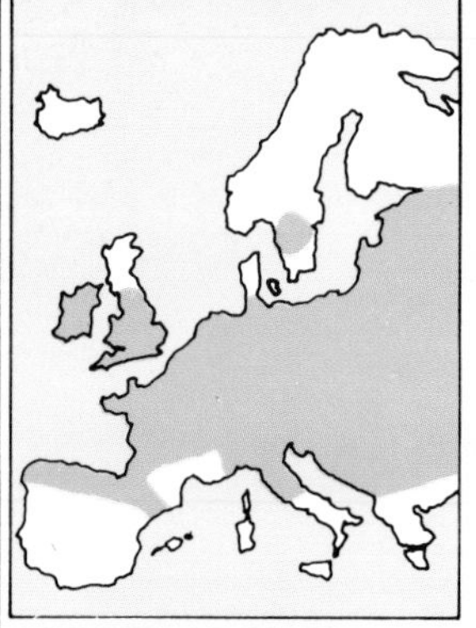

trrrrrrrrrrrrrr...

Song A high reeling trill, like a grasshopper.
Behaviour Sings from bushes and reedbeds, usually hidden but sometimes exposed. Often sings at night.
Habitat Marshy bushes, reedbeds and young conifers.
Nest Of grasses and leaves, on or close to the ground, hidden in a tussock. Six eggs, one or two broods.
Food Flies, moths, caterpillars, aphids and beetles.

The distinctive reeling song of the grasshopper warbler could easily be mistaken for that of a grasshopper itself, hence the name. The bird can be hard to locate since it throws its voice when singing, and in appearance – streaked brown upperparts, yellow-buff underparts and a rounded tail – it is hard to spot in the dense vegetation in which it breeds.

SAVI'S WARBLER *Locustella luscinoides*

Its unstreaked back and more rufous colour distinguishes this bird from its similar relative, the grasshopper warbler. Savi's warblers do not inhabit plantations and dry bushy areas, preferring marshy reedbeds, where their lower and more rapid song is often given from a prominent reed.

THIS CLOSE RELATIVE of the grasshopper warbler shows a greater affinity for reedbeds and fens – which has nearly been its undoing. Extensive drainage and reclamation of wetland areas, particularly in eastern England, meant that the Savi's warbler became extinct as a breeding species in Britain in 1856. It fared little better in Europe, as witnessed by its fragmented present-day distribution. It is a slim, rufous-brown bird, similar to but darker than a reed warbler. Its back and underparts are unstreaked and its tail is broad and rounded. There is a pale stripe above the eyes, and the whitish chin and throat merge with a buff upper breast. These warblers spend most of their time unseen in reedbeds, but will climb to the tops of reeds.

The song is very similar to a grasshopper warbler's, but can be distinguished with practice. It often begins with slow ticking notes and the trills, which are lower and have a buzzing quality, are generally shorter than a grasshopper warbler's, often lasting for less than 30 seconds. The singing bird may perch at the top of a reed or in a nearby bush, and turns its head as it sings, its whole body shivering. Song begins as soon as birds arrive at their breeding site and stops when breeding is under way. The call is a sharp *tswick*.

The nest, shaped like a deep cup, is built among the lower stems of reeds and sedges. Breeding begins in mid-April in southern Europe but does not start until late May in the north. Four or five eggs are normally laid and two broods can be reared. These birds feed on aquatic insects and larvae, and occasionally worms.

Savi's warblers breed from southern Spain and Greece north to West Germany and Poland. They are absent from many parts of central Europe, where their habitat no longer exists. They arrive in Europe from early April, departing again in August.

FACTS AND FEATURES

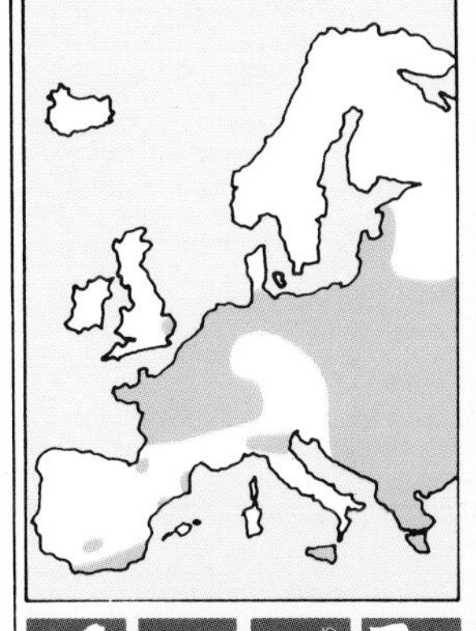

drrrrrrr...

Song A buzzing trill, lower and faster than a grasshopper warbler, and in shorter bursts.
Behaviour Sings from near tops of reeds and marshy bushes, often in full view. Sometimes sings at night.
Habitat Reedy swamps with scattered bushes.
Nest Of grass and leaves low among reeds and sedges. Four or five eggs, two broods.
Food Beetles, moths, flies, water insects and larvae, caterpillars and worms.

SEDGE WARBLER *Acrocephalus schoenobaenus*

A noisy and conspicuous summer visitor whose grating and churring song and calls can be heard from bushes and scrub near water. When perched the sedge warbler's pale buff stripe over the eye is clearly visible. In flight it shows a rusty-coloured rump. Its crown and back are streaked, while its underparts are a plain buff-white.

FACTS AND FEATURES

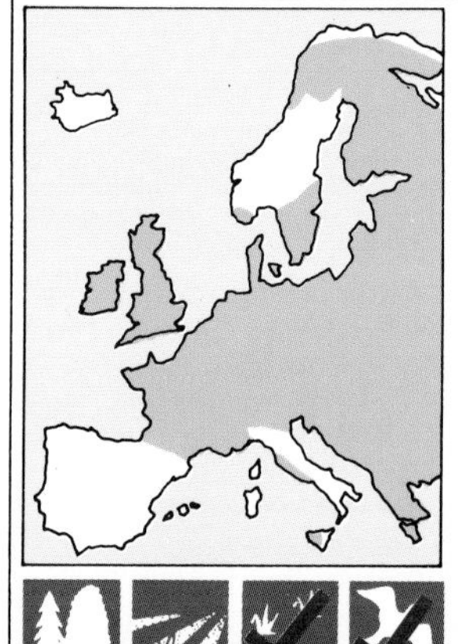

chit-chit-tuk-tuk-tuk-
chit-terwee-terwee...

Song A chattering series of repeated phrases containing musical and harsh notes.
Behaviour Sings from reeds and bushes and in song flight. Often sings at night.
Habitat Reedbeds, ditchside hedges and marshy vegetation.
Nest Of grass and moss in waterside vegetation, off the ground. Five or six eggs, one or two broods.
Food Flies, beetles, aphids, grubs and spiders.

IT IS DIFFICULT TO IGNORE the sedge warbler, whose noisy song is a feature of the summer across most of Europe. Its preferred habitat is reedbeds, bushes and dense vegetation along rivers, gravel pits and marshy ditches. But it is not as confined to wetlands as its relative, the reed warbler, being found also in dry woodland and occasionally cornfields. This active bird can creep rapidly through tangled vegetation and its flight is direct and low. It has pale brown, dark-streaked upperparts and whitish-buff underparts; its most distinctive feature is the broad, dark-bordered buff stripe above the eye.

The loud and varied song is full of trills, whistles and chatters with lots of *chits*. Notes are often repeated and their rich variety lends a somewhat erratic quality. A song may last a minute or more, often uttered from a conspicuous song post but more regularly from the edge of a bush. A song flight is also commonly seen, in which the bird rises almost vertically with a flapping flight, returning to its perch with tail spread and wings outstretched. This warbler's habit of singing at night has frequently resulted in it being identified as a nightingale, although their songs are dissimilar.

Sedge warblers are excellent mimics and often incorporate the songs and calls of swallow, reed bunting, blackbird, blue tit and house sparrow. Singing begins as soon as the birds arrive at their breeding grounds and continues until mid- or late July. The call note is a scolding *tuk*, which is often repeated as a rattle if the bird is excited.

The nests are built low in thick vegetation, and cuckoos occasionally lay in them. Five or six eggs are usually laid with some second broods.

Sedge warblers arrive in Europe from early April. They breed right up into northern Scandinavia, but are absent from much of the Mediterranean region. They depart in July, August and September to spend the winter in tropical and southern Africa. Their diet includes flies, midges, beetles and aphids.

MARSH WARBLER *Acrocephalus palustris*

THE MARSH WARBLER is perhaps the most accomplished European songbird, belying its rather dull appearance. It has an olive-brown back, yellowish-white underparts and a white throat. It often perches upright, but tends to move through vegetation with its tail held slightly down. This bird prefers dense cover and has shown a liking for osier beds with nettles and other rank rambling vegetation. It is a heavy-looking warbler, often appearing 'pot-bellied'. Generally secretive and hard to see, it may perch in the open in a tree or bush when singing.

The song is a lively warble, full of buzzes and trills. This warbler's exceptional imitative powers mean that no two songs are exactly alike, although most birds give a repeated nasal *za-weee* as part of the song. One individual may be able to mimic the songs and calls of up to 40 European species, while marsh warblers as a species have been recorded as imitating nearly 100 other birds, including greenfinch, goldfinch, yellowhammer, reed bunting, skylark, tree pipit and several other warblers. Phrases 'stolen' from European birds make up only part of the marsh warbler's repertoire; many others have been identified as belonging to species that it encounters on the wintering grounds in Africa.

The marsh warbler has a short song period, from early June to late July, and its singing is most intense as it sets up a territory and awaits a mate. The call note is a loud, repeated *tic* and in alarm it gives a churring rattle.

The nest is built one or two feet (less than one metre) above the ground, and is anchored to surrounding stems of willowherb or other plants with loops of nest material. Four or five eggs are laid and in such a short breeding season only one brood can be raised. Marsh warblers are rarely parasitized by cuckoos.

Marsh warblers are one of the latest migrants to return to Europe after the winter. A few arrive in the second half of May, with more birds in early June. They leave in August and stop in north-east Africa before moving farther south for their winter moult. They feed on flies, beetles, aphids, butterflies, moths, caterpillars and spiders, and some berries.

FACTS AND FEATURES

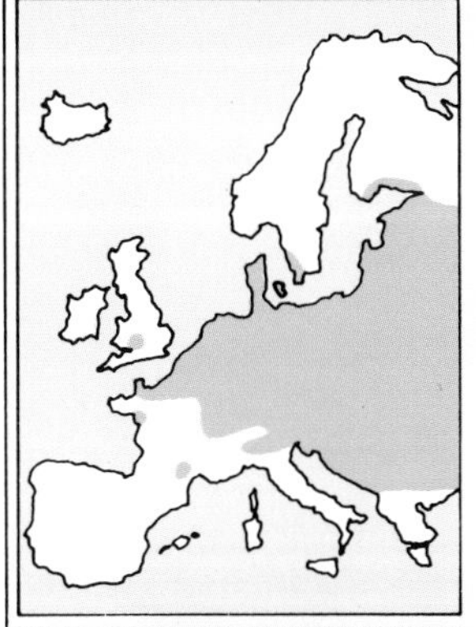

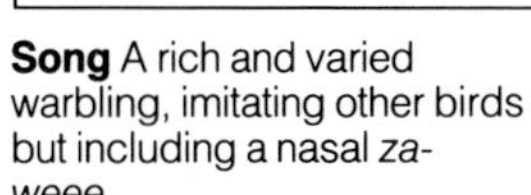

Song A rich and varied warbling, imitating other birds but including a nasal *za-weee*.
Behaviour Sings from tall plants, trees and bushes, sometimes at night.
Habitat Overgrown bushy areas with nettles, especially osier beds, often near water.
Nest Of grasses and fixed with looped 'handles' to surrounding vegetation. Four or five eggs, one brood.
Food Flies, damselflies, moths, beetles, aphids, caterpillars, spiders and berries.

Resembling a reed warbler but with olive-brown upperparts, pale underparts and a whitish throat, the marsh warbler is an unsurpassable mimic, able to imitate dozens of other birds' songs and calls. It is not so confined to reedbeds and is more likely to be seen at the edge of damp woodland in nettles, willowherb and brambles.

REED WARBLER *Acrocephalus scirpaceus*

SELDOM FOUND FAR from reeds and water, the reed warbler is typical of its group. Slim, round-tailed and with a flattish forehead, it differs from the marsh warbler in being a more rufous-brown on the back and rump. It is a shy and secretive bird, spending its time in the middle of reedbeds and other thick cover, and seen only briefly as it clings to a reed stem or flies across a patch of open water. Like its relatives, this warbler oftens shows itself best when in song. It favours *Phragmites* (common reed) but may also breed in other thick vegetation and bushes close to reedbeds. The diet includes flies, midges, damselflies, butterflies, moths, caterpillars, aphids and spiders.

The song can sometimes be difficult to distinguish from a sedge warbler's, particularly in spring when familiarity with the songs has not yet been re-established by the listener. The churring phrases have more uniformity and are slower and less lively than a sedge warbler's. Churring notes are repeated, often in threes, to form song sequences such as *kerr-kerr-kerr, kek-kek-kek, chirruc-chirruc-chirruc*. The favourite song perch is a reed, the bird clinging part of the way up or at the top, with beak wide open and throat fluffed out. Singing at night is fairly common and continues when daytime singing ceases, after pairs have formed. The song is heard from late April until the end of July, and may sometimes continue into August and even September. The alarm call is a low, scolding *churr*, prolonged and grating if the bird is excited.

The reed warbler is most likely to be seen perched on a reed stalk at the edge of a reedbed delivering its loud and repetitive song. With rufous upperparts, buff underparts and dark legs it is distinguished from the similar sedge warbler by its lack of streaking, thinner eye-stripe, and a song that has fewer whistles.

FACTS AND FEATURES

kerr-kerr-kerr, chirruc-chirruc-chirruc...

Song A varied warble of low phrases, each repeated two or three times and having a chirping quality.
Behaviour Sings when hidden in or perched on top of reeds and other vegetation. Often sings at night.
Habitat Reedbeds and bushes near water.
Nest Of grass and reed-flowers in reedbeds, anchored to surrounding stems. Three or four eggs, one or two broods.
Food Flies, damselflies, beetles, aphids, spiders and berries.

The nest is a cup of grass woven around several reed stems. Three or four eggs are usually laid and one or two broods reared. Reed warblers are one of the commonest host species to the cuckoo – in some colonies more than half their nests may be parasitized, resulting in either the nests being deserted or no young being raised apart from cuckoos.

Reed warblers arrive in Europe from mid-April until the end of May. They are found over most of Europe except for northern Scandinavia, Iceland, Ireland, most of Scotland, Wales and south-west England. They leave in August and September, when migrating birds may be seen away from their normal wetland habitat, in coastal bushes and trees. These warblers spend the winter in tropical Africa, where they are often heard singing in December and January.

GREAT REED WARBLER *Acrocephalus arundinaceus*

An extra-large version of the reed warbler with a voice to match, the great reed warbler can be tracked through a reedbed by the swaying of the reeds. In flight it has a pale rump, unlike the rufous rump of the reed warbler. The large nest which it builds suspended from reed stems is often host to the cuckoo.

A GIANT AMONG WARBLERS, nearly as big as a song thrush, the great reed warbler has a voice to match its size. It resembles a large version of the reed warbler, with its unstreaked rufous-brown back, long, stout bill and fairly long tail. Although secretive, it perches more out in the open, on trees, bushes and wires, compared with other warblers. Its size alone makes it distinctive and its progress through a reedbed can often be tracked by the movement of the stems. Like its smaller namesake it frequents beds of *Phragmites* (common reed) with open fresh water and trees or bushes nearby.

The song is loud, harsh and croaking, containing phrases like *kar-kar-kar*, *karra-karra-karra* and *gurk-gurk-gurk*. A single burst of song may be quite short or as long as 20 minutes without a break. This warbler does not imitate other birds, like its relatives. Song is given by the male, who perches on a reed stem, bush top or tree branch. His bill is held wide open as he delivers bursts of song. Males begin singing as soon as they arrive at their breeding site; they are heard from dawn to dusk, and occasionally at night. Singing continues until the female begins to incubate. Call notes are a harsh *chack* and a croaking *churr* of alarm.

The nest is built around reed stems, which occasionally collapse under the weight, especially after the young have hatched. It is anchored to the stems by wet material which is woven around them and allowed to dry. The nest is usually positioned about three feet (one metre) above the water, well hidden in the reeds, but nevertheless sometimes parasitized by cuckoos. A large reedbed may contain several pairs, nesting amicably together in fairly close proximity. Great reed warblers may also breed in thick vegetation beside streams and ditches. From four to six eggs are usually laid and only one brood is raised.

Great reed warblers breed over most of Europe, with the exception of Britain and Ireland, Iceland and most of Scandinavia. They arrive from mid-April to early May, leaving in August and September to winter in tropical and southern Africa, where they may sing occasionally. These large birds eat dragonflies, damselflies, mayflies, aphids, caterpillars and grubs, and they may feed from the ground, when they can resemble thrushes.

FACTS AND FEATURES

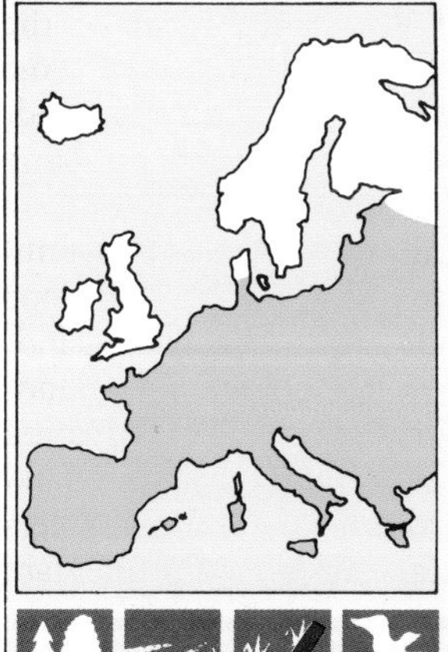

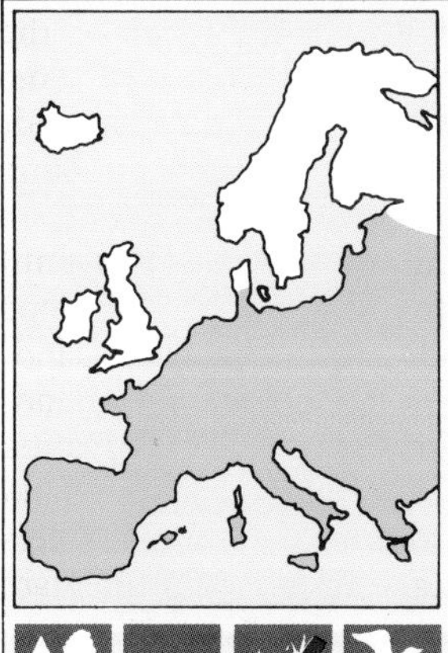

kar-kar-kar, karra-karra-karra,
gurk-gurk-gurk

Song Loud, harsh and croaking phrases with typical notes *karra* and *gurk*.
Behaviour Sings from reeds and nearby bush tops.
Habitat Reedbeds near open water.
Nest Of leaves and roots among reed stems. Four to six eggs, one brood.
Food Flies, dragonflies, aphids, caterpillars and spiders.

ICTERINE WARBLER *Hippolais icterina*

THE ICTERINE WARBLER, and its close relative the melodious warbler, belong to a family of heavy-bodied, square-tailed, rather clumsy warblers that can be difficult to distinguish without hearing their songs. They lack the grace and charm of willow warblers and tend to crash through the trees and bushes they feed in. The long bill, low, sloping forehead and 'peaked' look of the crown are good identification features.

In spring, the icterine warbler is olive-green above and lemon-yellow below, although some individuals are greyer above and whitish below. Autumn plumage is brown or grey above with yellow-white underparts. The long wings project beyond the upper tail coverts when closed, and pale edges to the wing feathers form a light wing panel. The species inhabits deciduous woodland, parkland, orchards and gardens and in parts of Europe it is a common garden bird.

The song is a loud, rapid warbling. The varied phrases often contain imitations of other birds such as skylark, goldfinch, starling (itself a mimic) and tawny owl. The icterine warbler's own phrases are repeated whistles, chatters and churrs such as *wee-choo, choo-choo-choo-chit, chee-chee-wee*. A short phrase which recurs is a musical *dideroy*. Singing is from tree branches and bushes, sometimes from telegraph wires and rarely in flight. Occasional nocturnal singing also occurs. The song period is from early May to early July. Call notes are a sharp *bik, bik* or *tek, tek* and a churring of alarm.

The nest is built in the fork of a tree or bush, especially a fruit tree. It is often secured to the branches with loops of grass. Four or five eggs are laid and only one brood is reared.

Icterine warblers are absent from southern Europe, most of western Europe and northern Scandinavia. They arrive on their breeding grounds from mid-April to May. Although not a breeding species in Britain, birds occasionally appear in spring. The southward migration begins in July and continues to September, and in autumn the icterine warbler is a regular visitor to the east and south coasts of Britain and Ireland. The birds spend the winter in Africa. Insects and their larvae make up the diet, with berries in autumn.

FACTS AND FEATURES

wee-choo, choo-choo-choo-chit, dideroy

Song A loud, rapid warbling of varied phrases with chatters and churrs.
Behaviour Sings from a high song post on a tree and also from cover.
Habitat Gardens, parks and woods.
Nest Of grass, leaves, wool and spiders' webs in the fork of a tree or bush. Four or five eggs, one brood.
Food Flies, beetles, moths, aphids, spiders and berries.

Resembling a large stout-billed willow warbler, the icterine warbler has yellow underparts and a green-brown back. It also has a peaked crown and a distinctive pale base to its bill. Pale edgings to the long wing feathers form a panel on the closed wing which helps to distinguish it from its other similar cousin, the melodious warbler, but it can be distinguished from other warblers chiefly by its song – repeated whistles, chatters and churrs. The call notes are a sharp bik, bik *or* tek, tek.

MELODIOUS WARBLER *Hippolais polyglotta*

FACTS AND FEATURES

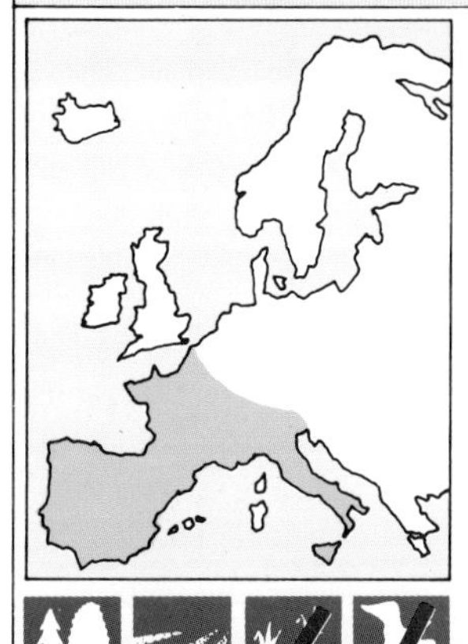

chizzick, whit-tchu, whi-whi-tchu

Song A hurried musical warble, highly imitative and containing sparrow-like phrases.
Behaviour Sings from the cover of bushes, occasionally an exposed perch.
Habitat Woodland, hedges and bushes, often near water.
Nest Of grass and vegetable down in the fork of a bush. Four eggs, one or two broods.
Food Flies, ants, beetles, grasshoppers and berries.

SIMILAR IN APPEARANCE to the icterine warbler, the melodious warbler can be distinguished by its lack of a pale wing panel and its shorter wings, which do not project as far when closed, giving a somewhat fluttering flight. It is heavily built and rather clumsy in its movements as it feeds in bushes and shrubs. This warbler is found in woodland but not in parks and gardens, preferring denser shrubs or hedges of gorse, bramble, hawthorn or tamarisk, often near water.

The song is a hurried warble with many sparrow-like chattering phrases, often introduced with passages like *chizzick, whit-tchu, whi-whi-tchu*. Some imitative phrases can also be included, but not as many as the icterine warbler. Exposed song posts may be used, but the melodious warbler is generally quite secretive, singing from the centre of a bush. Singing in flight is fairly frequent and the bird often moves to the top of a bush in stages while continuing to sing. Song period is from May to July. The calls are a sharp *tic* and a chirping chatter.

The nest is built in the fork of a shrub, often quite low down, and usually woven around the branches. Four eggs are laid and two broods sometimes raised.

Melodious warblers breed mainly in south-western Europe, in France, Switzerland, Italy, Spain and Portugal. In spring, birds begin to arrive at the beginning of May; they return to their wintering quarters, in tropical West Africa, in late July and August. They are seen in other countries as rare breeders or visitors. In Britain, they mainly occur along the south coast of England in August. The main food items are insects like flies, beetles, ants and grasshoppers as well as their larvae, and berries are also taken in autumn.

Two other related species breed in Europe – the olivaceous warbler, *Hippolais pallida*, and the olive-tree warbler, *Hippolais olivetorum*. The former is pale brown, resembling a marsh or reed warbler, and has a song reminiscent of a sedge warbler but containing a repeated *tchick* note. The latter is large and grey with a pale wing panel, and a slow warble with *took* and *tchoo* notes. Both are found in southern Europe around the Mediterranean.

The shorter plain wings (and consequent fluttering flight) and its brown, rather than grey, legs help to distinguish the melodious warbler from the similar icterine warbler. Its song is faster and more chattering than an icterine. Like all members of its family, it feeds clumsily, crashing about through bushes and scrub.

DARTFORD WARBLER *Sylvia undata*

An inhabitant of shrubby heathland, especially gorse and heather, the Dartford warbler appears dark grey, but the underparts of the male are purplish-brown and at close quarters a red eye-ring is also visible. The Dartford warbler's long tail is often cocked.

FACTS AND FEATURES

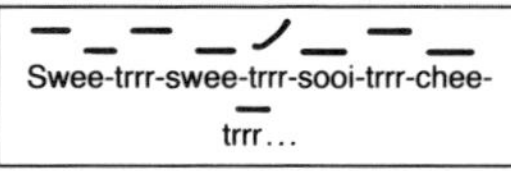

Song A rapid warble interspersed with churring notes.
Behaviour Sings from the top of scrub and in song flight.
Habitat Heaths and downland with gorse.
Nest Of grass, heather and wool in gorse or long heather. Three or four eggs, one or two broods.
Food Beetles, butterflies, caterpillars, flies and spiders.

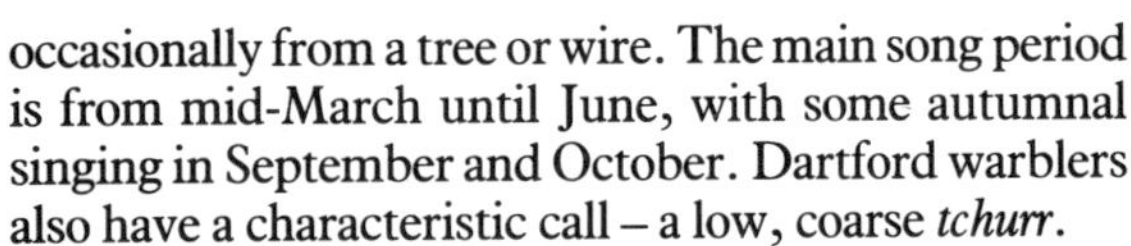

ORIGINALLY NAMED AFTER the town in Kent, the Dartford warbler is rare in Britain today. A small, long-tailed and very dark-coloured warbler, it is usually only seen as it flits between patches of scrub with an undulating flight, occasionally alighting on the top of a brush of gorse or sprig of heather. The long tail is often flicked up or held cocked, and if alarmed, this warbler can raise its crown feathers to give its head a peaked appearance. The upperparts are a mixture of slate grey and dark brown and the underparts are pinky-purple; a close view may show the orange-red eye-ring.

The song is a lively, rapid warble with churring notes, characteristic of most scrub-dwelling warblers. Each burst of song is short, made up of high notes followed by a churring *trrr*, for example, *swee-trrr-swee-trrr-sooi-trrr-chee-trrr*. Singing is from a perch at the top of a gorse clump, or in a vertical song flight, or occasionally from a tree or wire. The main song period is from mid-March until June, with some autumnal singing in September and October. Dartford warblers also have a characteristic call – a low, coarse *tchurr*.

Nests are built in gorse, heather and scrub, about one foot (30 centimetres) from the ground. The male builds a number of flimsy versions that are not used, the female being responsible for the breeding nest. Four eggs are laid and early breeders usually manage a second brood.

Dartford warblers are resident throughout their range. They are found in southern and western Europe, in thorny scrub; in Britain, they are confined to the thickly-vegetated southern heaths of Dorset, Surrey and Devon. Like all resident insect-eaters they suffer badly in hard winters. Thick snow, together with freezing temperatures for a prolonged period, result in many deaths. After the severe winter of 1962-63 an estimated 11 pairs survived, but numbers had built up to more than 400 pairs in 1984. Occasionally continental birds are recorded in Britain, and there may be some partial migration of British birds in winter. These warblers feed on beetles, flies, caterpillars and spiders.

A closely related species which has an even more confined range in Europe is the Marmora's warbler, *Sylvia sarda*. It is found only in north-eastern Spain, the Balearic islands, Corsica, Sardinia and Sicily. It has a grey back and underparts and a faster, higher song.

SUBALPINE WARBLER *Sylvia cantillans*

FACTS AND FEATURES

witchatrrr-seetrrr-peetrrr...

Song A sweet warbling chatter with higher churring notes than a Dartford warbler.
Behaviour Sings from hidden perch in bushes and trees, sometimes in song flight.
Habitat Open woodland, bushes and thickets.
Nest Of grass and thistledown in a low bush. Three or four eggs, two broods.
Food Flies, beetles, crickets, spiders, caterpillars and grubs.

Like so many of the scrub warblers, the subalpine warbler shows a grey back and dark tail with white outer feathers. The male, however, is unmistakable with orange underparts and a clear white moustache. Its wing feathers have brown edgings and it has an orange eye-ring. Females have more buff-coloured underparts.

THIS MEDITERRANEAN SCRUB warbler is perhaps the most attractive of all the *Sylvia* warblers. It resembles a pale Dartford warbler but has a shorter, white-edged tail. The back is grey, the throat and upper breast are chestnut-orange, and a distinctive white moustache runs from the base of its bill. Orange legs and a bright orange eye-ring can be seen at close quarters. The subalpine warbler inhabits scrub with scattered bushes and trees, skulking in the thick of the vegetation and occasionally dashing with a low, fast flight into the next bush. Holm oaks, junipers and olive trees provide it with insect food, while broom, cistus, myrtle, gorse and other shrubs furnish nesting sites. When excited it cocks its tail and raises its crown feathers.

The fast, chattering warble is similar to the song of a Dartford warbler, but each burst lasts longer and the notes are sweeter with higher churrs, such as *witchatrrr-seetrrr-peetrrr-weetrrr-sittitrrr*... Songs are usually delivered from a partly concealed tree branch or the edge of a bush, or occasionally in the open at the top of a tree or bush. This warbler also has a whitethroat-like song flight, fluttering into the air and descending to a bush. The song period is from early April to late June. Its calls are a sharp *tec* and a *chat-chat-chat-chat* of alarm.

Nests are constructed about one to three feet (30 to 90 centimetres) from the ground. The nest is a neat cup with three or four eggs and one or two broods.

Subalpine warblers breed in all European countries bordering the Mediterranean. They arrive in Europe from late March onwards, leaving in late August and September to winter in West Africa south of the Sahara, in areas of acacia scrub. This species is a vagrant to Britain, in both spring and autumn, usually on the south and east coasts but occasionally as far north as Fair Isle. The main food is insects and their larvae, including flies, beetles, crickets, stick insects, caterpillars, grubs and spiders.

Another Mediterranean species which resembles the subalpine warbler is the spectacled warbler, *Sylvia conspicillata*. More confined to scrub, its song is a high and scratchy warble, *tchirrit-it-it-urritt-chit-chit*, often given in song flight. In appearance it is like a pale subalpine warbler with a white throat, rusty wings and white eye-ring.

SARDINIAN WARBLER *Sylvia sarda*

A resident of the Mediterranean, the Sardinian warbler with its grey back and pale underparts is less colourful than some of its relatives. The black head of the male, contrasting with a white throat, makes it readily identifiable. They are noisy birds, always giving rattling calls from the scrub and bushes where they live. The Sardinian warbler can be hard to see, keeping to dense vegetation, but its loud calling indicates its presence.

THIS IS THE COMMONEST and most noticeable of the Mediterranean scrub warblers. At a brief glance it might be mistaken for a male blackcap with its black crown and greyish back. A closer look reveals that the black extends below the eye and covers all of the head, contrasting with the white throat. In flight, the Sardinian warbler's dark tail shows its bright white edges. This bird is found in low scrub, thickets and bushes of gorse, juniper, bramble and cistus. While still secretive, it is very active and hops to the top of a shrub to look around, or creeps around the edge of a bush instead of through the centre. Like its relatives, it cocks its tail and raises its crown feathers when excited.

It may be virtually impossible to ignore the presence of a Sardinian warbler, because of its loud and frequent calling. One of the most noticeable calls is a sharp, rattling *treeka-treeka-treek* and there is also a stuttering *stititititititic* of alarm. Both calls may be incorporated into its song, which is a fast, persistent, chattering warble. The song phrases are broken by churring, rattling notes: *seeoo-chrr-pipichrrr-chrr-piseeoochrr*. Songs are delivered from the top of a bush or low tree, and also in song flight, in which the male flies into the air and descends to a nearby bush while singing. In spring a patch of scrub can seem alive with these warblers, all singing and chattering at once.

The nest is built low in scrub or in a bush. Normally three or four eggs are laid and two broods are raised.

Sardinian warblers breed along the Mediterranean coast from Portugal to Greece. Some individuals migrate south in winter to North Africa or even farther, while others remain. They have occurred as rare vagrants to Britain in spring and autumn, often remaining for weeks in a suitable habitat. The main diet is flies, aphids, caterpillars and spiders, although in the autumn they eat fruits such as figs and grapes.

Ruppell's warbler, *Sylvia rueppelli*, looks like a Sardinian warbler, but with a black throat and white moustache. It is found in southern Greece and the Greek islands; its song is similar to a Sardinian warbler's but more musical.

FACTS AND FEATURES

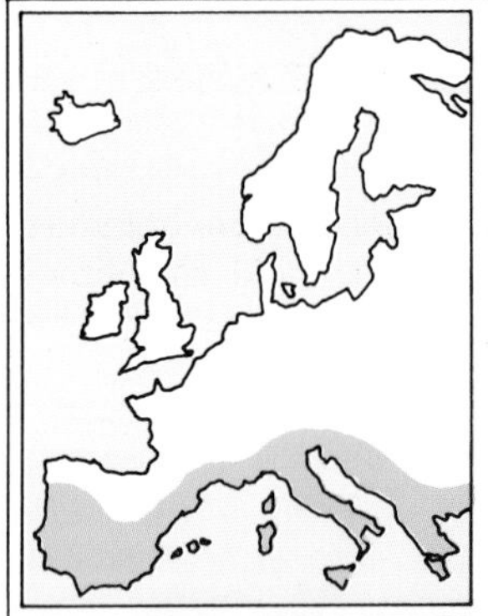

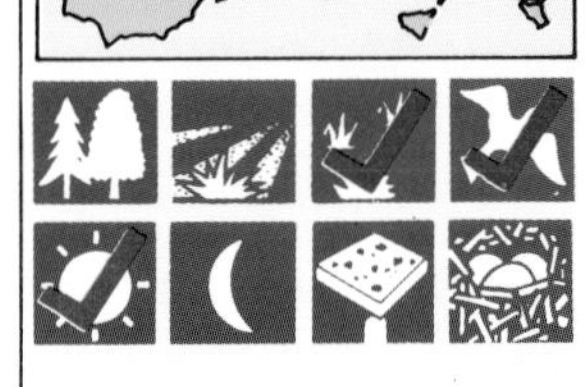

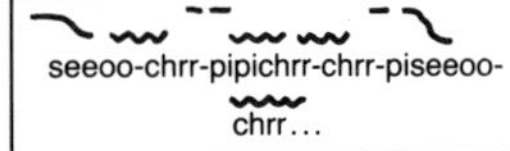

Song A loud and fast warbling with phrases broken by churring, rattling notes.
Behaviour Sings from cover, an exposed perch or in song flight.
Habitat Scrub, thickets and dense undergrowth.
Nest Of grass and down in a bush or scrubby vegetation. Three or four eggs, two broods.
Food Flies, aphids, caterpillars and spiders; fruits in autumn.

LESSER WHITETHROAT *Sylvia curruca*

THIS OTHERWISE ordinary-looking grey warbler is distinguished by a dark patch on its ear coverts, giving it a shrike-like face. Dark legs, lack of a white eye-ring and a short tail also help identification. In flight it reveals white outer tail feathers, like many of the *Sylvia* warblers. The lesser whitethroat is an active bird, always on the move through bushes and hedges, which it seems to frequent more than its relative the whitethroat, the latter favouring brambles.

The lesser whitethroat's song is one of the least inspiring of all the *Sylvia* warblers and is best described as a loud rattle, *chicker-chicker-chicker*..., rather like a cirl bunting. It may be preceded by a quiet sub-song which sounds more like a typical warble and gives the impression that the bird is singing softly to itself. This sub-song can only be heard when close to the bird, which may suddenly break into the rattle that can be heard several hundred yards (metres) away. The song is usually given from cover, often while the bird is moving. Males begin singing as soon as they find a suitable habitat, and on migration they even sing in gardens and parks, before they have settled down. The song period is from late April to early July. This whitethroat's alarm call is a hard *tac-tac*, similar to a blackcap, and a harsh *churr*.

The nest is built deep in a hedge or scrub, with a preference for blackthorn, hawthorn, rose or bramble. It is positioned between branches and sometimes secured to them. Between four and six eggs are laid and one brood is normally produced. The lesser whitethroat's diet consists of flies, beetles, aphids, ants, moths, butterflies and caterpillars.

Food is usually picked off leaves and branches but may be caught in flight. Fruit, such as blackberries and raspberries, is occasionally eaten in the autumn.

Lesser whitethroats are summer visitors to Europe. In Britain they are found mainly in southern England, with fewer in the north and west; they begin to arrive in late April and early May, and start to return in August and September. They are rare in Scotland and Wales and seldom reach Ireland. In Europe, these birds are absent from western and southern France, Spain and Portugal. This eastern distribution is partly explained by their wintering grounds, which are in north-east Africa, especially Ethiopia. They migrate south-east in the autumn, probably via Israel and Egypt.

FACTS AND FEATURES

whhao-ti-worr, chicker-chicker-chicker...

Song A loud rattle, often preceded by a quiet warble.
Behaviour Sings from cover in a bush, hedge or tree.
Habitat Thick hedgerows, tall scrub and shrubberies.
Nest Of grass, stalks and roots in bramble or hawthorn. Four to six eggs, one brood.
Food Flies, beetles, ants, caterpillars, aphids and berries.

A grey head and dark cheeks give the lesser whitethroat a shrike-like appearance. It has white underparts and a grey-brown back with plain wings, making it easily distinguishable from the whitethroat. It is found in thicker scrub and hedges more often than the whitethroat and has a loud rattling song.

WHITETHROAT *Sylvia communis*

LOCAL COUNTRY NAMES for birds are often extremely apt – none more so than 'nettle creeper' when describing a whitethroat. It is a bird of scrub, bushes and hedges, rarely venturing into trees; while it is less shy than many other warblers, it still tends to skulk in undergrowth, its slim form slipping easily between the branches and leaves. During migration the whitethroat may be seen in gardens and parks, but it is only likely to breed in the most rural of gardens. It is usually only seen flying from one patch of brambles or nettles to the next, identified by its white-edged tail, slightly rufous wings and jerky flight.

Whitethroats were once common in roadside hedgerows, but with the destruction of many hedges its numbers have decreased. In 1969 a dramatic crash in whitethroat numbers was detected – more than three-quarters of the breeding population failed to return from their African wintering grounds. Drought conditions there were thought to be the cause, and there has been no real increase in numbers since.

The lively and scratchy song is made up of 10 to 12 notes, *cheechiwee-cheechiweechoo-chiwichoo*. It is usually given from dense cover, in the thick of brambles or a hawthorn hedge. In contrast there is a conspicuous song flight in which the male ascends singing into the air, sometimes as high as 30 feet (about 10 metres), and descends with a dancing motion to his original bush or another nearby. Whitethroats sing from mid-April to the end of July. Their presence is also announced by a harsh, grating *tcharr* call, often used if a bird is disturbed near its nest, or a *tac-tac* of alarm.

FACTS AND FEATURES

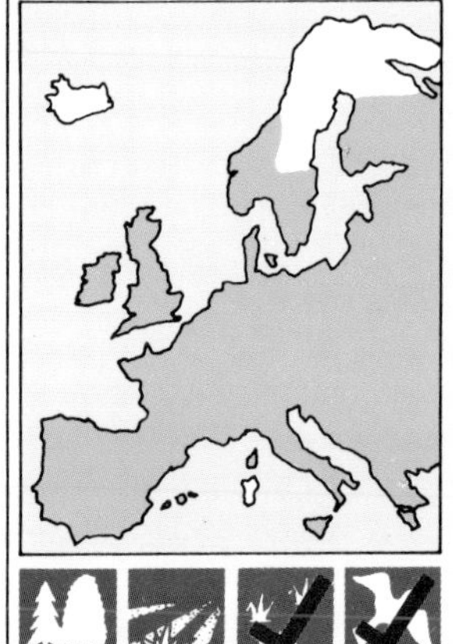

Cheechiwee-cheechiweechoo-chiwichoo

Song A short, lively warble.
Behaviour Sings from the cover of a hedge or bush and also in conspicuous song flight.
Habitat Open hedges and scrubby bushes.
Nest Of grass and roots in low bramble or other bush. Four to six eggs, two broods.
Food Beetles, moths, caterpillars, flies, ants, spiders and berries.

Nests are invariably built close to the ground in scrub and hedges, especially brambles, nettles, grasses, hawthorn, gorse and blackthorn. Four to six eggs are laid and two broods may be reared.

Whitethroats arrive in Britain from early April to late May, and leave from the middle of July to spend winter in tropical and southern Africa. They eat mainly insects such as beetles, moths and their caterpillars, ants, flies, aphids and also spiders.

The male whitethroat has a grey head, contrasting with its brown back; females have a browner head, but both have a white throat and a rufous patch on the wings. The lively warbling song of the whitethroat is often given in flight. Whitethroats will skulk in the middle of brambles, announcing their presence only with a low, scolding tcharr.

GARDEN WARBLER *Sylvia borin*

FACTS AND FEATURES

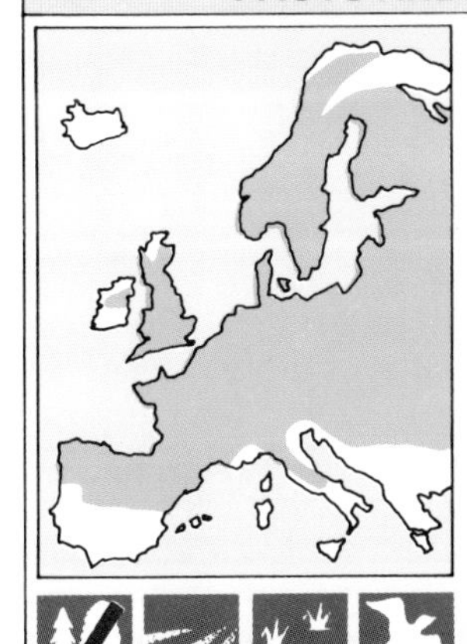

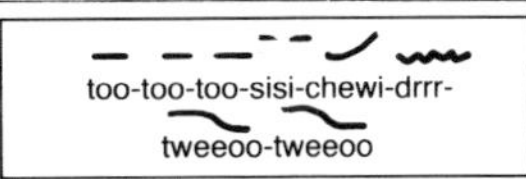

Song A long mellow warbling, more rapid and even than that of the blackcap.
Behaviour Sings from the cover of trees and bushes, often while moving about.
Habitat Deciduous woodland and hedgerows.
Nest Of grass, moss and twigs in shrubs and brambles. Four or five eggs, one brood.
Food Caterpillars, beetles, aphids, spiders and berries.

POSSIBLY THE MOST nondescript of all British birds, the garden warbler's song makes up for its drab appearance. It is a plump warbler, brown-grey in colour with paler underparts and a light eye-ring. It inhabits woodland, bushes and scrub and despite its name it is not a common garden bird except in large, rather overgrown gardens.

The song can at first be difficult to distinguish from a blackcap's. However, a comparison of the two soon reveals that the garden warbler's is more mellow and less varied, and has a more even volume and form, *too-too-too-sisi-chewi-drrr-tweeoo-tweeoo*..., lacking the loud, fluty outbursts of the blackcap. Song may be delivered from low in brambles and other scrub or high in the tree canopy. Sometimes the bird is stationary, while at other times it moves around, occasionally singing in flight. Song commences as soon as males arrive in a suitable breeding habitat and continues vigorously until they pair, when they sing less enthusiastically until mid-July. The commonest call note is a firm *teck-teck* and there is also a scolding whitethroat-like *tchurrr*.

The male may build a number of simple nests and then entice a female to inspect them. Often none of these is used, but occasionally she selects one and completes it. Nests are built fairly close to the ground in brambles, nettles, hawthorn and other scrub; deciduous woodland is preferred, but these birds also nest in young conifer plantations. Four or five eggs are laid and only one brood is reared.

Like most warblers, garden warblers are summer visitors to Europe and arrive in late April and May. They are found over most of the region except for around the Mediterranean, and are rare in northern Scotland and Ireland. They leave from August onwards and spend the winter in Africa.

Garden warblers feed on caterpillars, flies, beetles, aphids and spiders during the spring and summer. However, they eat large amounts of many wild and garden berries in the autumn.

The garden warbler is a plain-looking bird distinguished only by its lack of interesting features! It has brown upperparts and buff underparts, and grey legs. At close range a faint pale eye-ring can just be seen. In size and shape it resembles a blackcap, but usually arrives later in the spring and its song is mellower.

BLACKCAP *Sylvia atricapilla*

FACTS AND FEATURES

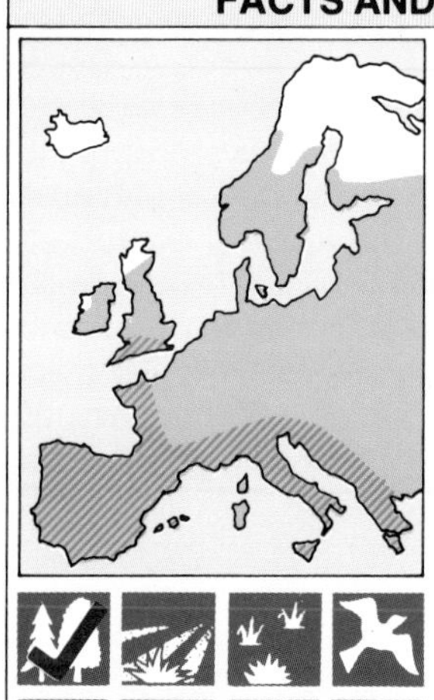

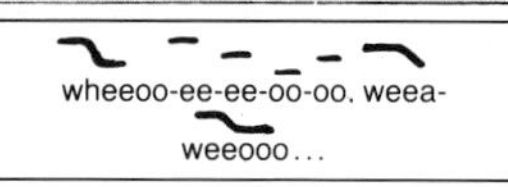

Song A rich warbling with wide-ranging, deliberate, fluty notes.
Behaviour Sings from tree canopies and tall bushes.
Habitat Deciduous and mixed woodland, coppices and shrubberies.
Nest Of grass and roots in a low bush or shrub. Five eggs, one or two broods.
Food Beetles, caterpillars, flies, aphids and berries. Wintering birds take fat, fruit and other scraps from a bird table.

THE BLACKCAP'S SINGING prowess has justly earned it the nickname 'northern nightingale'. It is similar in coloration to the garden warbler, but the male has a jet-black crown and the female a red-brown one. It is found in mature mixed or deciduous woodland, tall hedges, thickets and some shrubs.

The song is a rich warbling that contains purer notes than that of a garden warbler. Its main characteristics are loud, fluty whistling phrases that are delivered vigorously after the initial quieter warbling. Notes are uneven in pitch, rising and falling to form a musical phrase, ... *wheeoo-ee-ee-oo-oo, weea-weeoo*.... The song phrases are sung in rapid succession, the bird often perching motionless on a branch, its bill open and throat fluffed out, but it may also sing when moving about. Most song is given from the tree canopy or tall bushes. Blackcaps begin to sing as soon as they arrive in early April, although over-wintering birds may sing earlier than this, even in midwinter. The song continues, more quietly after pairing, until the beginning of July. In autumn these warblers frequently sing a high and restrained sub-song, which may also be used as a prelude to full song in spring. The common call note is a hard *tac-tac* similar to a lesser whitethroat.

The male may construct simple nests before the female arrives, but she builds the actual nest, usually only a few feet from the ground in a shrub, low bush or tree. Five eggs are laid and two broods can be reared in southern England.

Blackcaps are both summer and winter visitors to Britain and Ireland. Our breeding birds arrive in early April and depart in August and September. In autumn, they and other birds from western Europe migrate through Spain to North and West Africa, while those from the east migrate via the Middle East. Some continental birds do not migrate so far and spend the winter in Britain.

In summer, these warblers feed on caterpillars, flies, aphids and beetles; in late summer and autumn, they eat berries and other fruits; and in winter, they will come to bird tables where they like bread, fat and fruits but also eat peanuts, cheese and seeds.

The smart glossy black crown of the male and the reddish crown of the female distinguish the blackcap from any other woodland warbler. Its chief attraction, however, is its rich warbling song. It is a common woodland bird through most of Europe. Small numbers spend the winter in Britain and may visit gardens for food.

BONELLI'S WARBLER *Phylloscopus bonelli*

THIS DELIGHTFUL, SMALL, active warbler is typical of the tree-dwelling leaf warblers. About the size of a willow warbler, it is distinguished by its white underparts, greyish head with indistinct supercilium ('eyebrow'), yellowish rump and yellow edgings to the wing feathers. Its pale head and dark eye lend a distinctive facial expression. Always restless, it flits through the tree canopy, picking off insects such as caterpillars and occasionally darting out to catch a fly or other passing insect. It inhabits both broad-leaved and coniferous woodland, including oak, birch, poplar, alder, beech, pine, larch and spruce. Woodlands along riversides, on sunny hillsides and especially in more mountainous regions are the most suitable.

The song is a trill of about nine to 11 notes, lower than a wood warbler's and lacking any acceleration or introductory notes. It is a consistent and evenly-pitched song and could be likened to a high-pitched greenfinch trill. The bird sings as it moves around among leaves and branches. Song begins in April, on arrival in southern Europe. The call is a willow warbler-like *hooeet*, and eastern birds give a *chup* or *cheep*.

The nest is built on the ground, in a bank or depression among leaf litter and vegetation. From four to six eggs are laid and breeding commences in early or mid-May, resulting in only one brood.

Most Bonelli's warblers breed in southern Europe, extending across the Mediterranean from Spain to Italy, with some reaching as far north as Holland and East Germany. The western birds spend winter in tropical West Africa; birds that breed in the Balkans and Near East belong to a separate sub-species and winter in north-east Africa. Migration southwards takes place in August and early September. The species occurs in Britain as a rare vagrant, usually in autumn, although occasionally singing birds have arrived in spring.

FACTS AND FEATURES

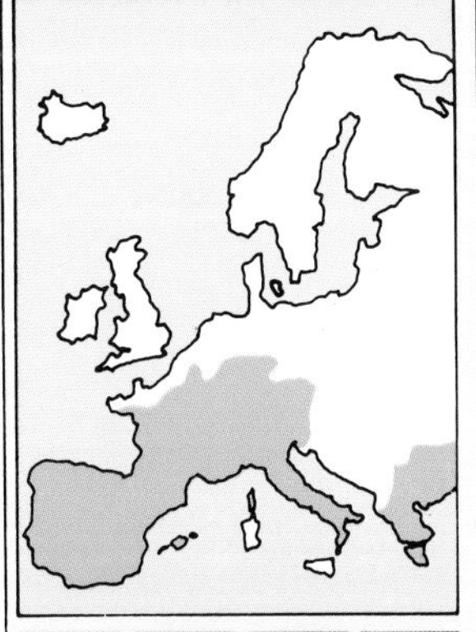

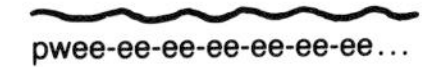

Song A short, even trill, similar to the trill of a greenfinch.
Behaviour Sings from the tree canopy.
Habitat Deciduous and coniferous woodland.
Nest Of grass, domed with a side entrance, on the ground or a bank. Four to six eggs, one brood.
Food Flies, caterpillars and other insects.

Like all the leaf warblers, Bonelli's warbler can be identified by its trilling song, which differs from a wood warbler in being lower in pitch. It has a greyish-green head, and greenish-brown back with white underparts. Its rump is yellowish and there are yellow edgings to the wing feathers. A woodland bird, it tends to stay high in trees to feed, coming to the ground only to breed.

WOOD WARBLER *Phylloscopus sibilatrix*

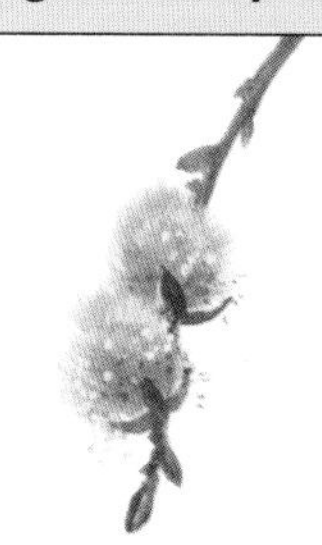

One of the brightest-coloured leaf warblers, the adult wood warbler has a green back, white belly, yellow breast and a broad yellow eye-stripe. The wing feathers are also yellow-edged. The shivering trill of the wood warbler's song carries some distance through woodland. The wood warbler feeds in trees and will come close to the ground, especially when it is giving its mating display flight.

FACTS AND FEATURES

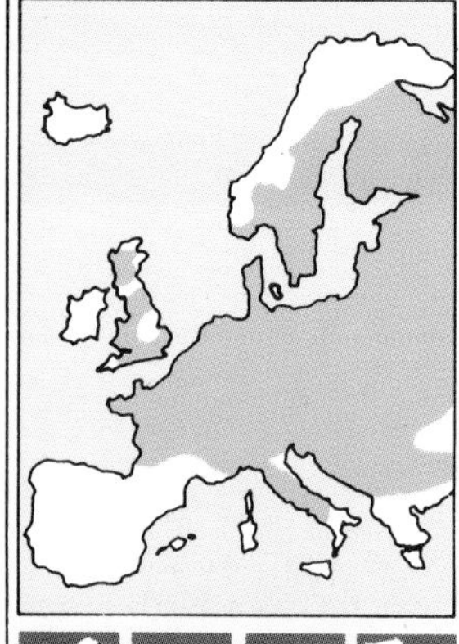

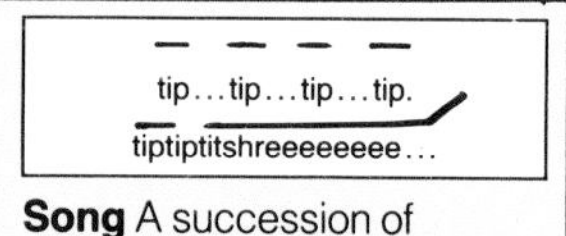

Song A succession of accelerating *tip-tip-tip...* notes followed by a trill. Also a series of piping notes, *pew-pew-pew-pew*.
Behaviour Song given from a perch or in display flight.
Habitat Mature open deciduous woodland of oak, beech, chestnut or birch.
Nest Of bracken, leaves and grass, domed with a side entrance, hidden in grass. Six or seven eggs, one brood.
Food Caterpillars, flies, moths, beetles and aphids.

THE LARGEST and most brightly coloured of the European leaf warblers, the wood warbler is also the most arboreal (tree-dwelling). It lives in open deciduous woodland, particularly sessile oak, but also beech, chestnut and birch. Cover beneath the trees is a habitat requirement, with some bramble or bracken but not thick bushes. This bird has yellowish-green upperparts, a yellow throat contrasting with the white belly, and a distinctive broad yellow supercilium ('eyebrow'). When feeding, it holds its body horizontally and may droop its wings slightly, sometimes flying out to catch a midge or other airborne insect.

Since wood warblers spend a great deal of time in the leaf canopy, it is not always easy to see them – however, it is difficult to ignore them in song, and their loud trill can be heard some distance away. The song begins as a slow repetition of a short note, *tip* or *sip*, which rapidly accelerates into a cascading trill, *tip... tip... tip.. tip tipti, ptitshreeeeeeee...*, the whole phrase lasting from two to five seconds. The song is usually delivered from a tree branch, and the perched bird quivers its wings and tail when trilling. The male also has a song flight which he performs in front of the female. He launches himself upwards at the beginning of the phrase and spirals down with rapid, shallow wingbeats to land near his mate. A second type of song consists of a series of plaintive *pew* notes, which may be uttered between bursts of the trilling song phrase. Singing begins in early May and continues until mid-July. The typical call note is similar to the *pew* used when singing, but less ringing.

The domed nest is built on the woodland floor, hidden in leaves, grass or other vegetation. Six or seven eggs are laid, and because of the bird's late arrival in Europe, only one brood is produced.

Wood warblers arrive back in England from mid- to late April onwards. They are rarely encountered before reaching a suitable breeding habitat. Departure begins in late July and continues until the end of August or early September. These warblers spend the winter in tropical Africa, from north-east Kenya to the Ivory Coast. They feed on insects such as caterpillars, midges, flies, moths, beetles and aphids – picking them off the foliage, fly-catching and also hovering to get at the more inaccessible insects.

CHIFFCHAFF *Phylloscopus collybita*

THIS SMALL WARBLER is the first of its family to arrive back after the winter, and its distinctive song, from which it derives its name, is the first to brighten a sunny March day. A slim, active bird, it spends its time in the leaf canopy, hopping from branch to branch and often flicking its wings and wagging its tail. It also feeds on the ground, moving around with short hops. These warblers are methodical in feeding, carefully inspecting twigs and leaves, or darting out to catch a fly, or hovering to pick a caterpillar from a leaf.

The chiffchaff is dull olive-brown above and buff below, with a pale 'eyebrow', looking altogether dingier, and with darker legs, than the willow warbler. In autumn, young birds are much yellower, both above and below. This warbler lives in deciduous and mixed woodland with good ground vegetation, and in commons, parks and tall hedgerows; in autumn and winter they may visit gardens.

On a warm spring morning the chiffchaff may be found perched on a high branch of a pussy willow, singing its repetitive song – a simple *chiff-chaff-chiff-chiff-chaff*..., with a varying number of *chiffs* and *chaffs*, the former pitched noticeably higher than the latter. Bursts of song may last from a few seconds to more than 20 seconds. Singing commences in late March, although wintering birds may sing earlier, and continues until late July, occasionally being heard in September. The commonest call note is a plaintive *hooeet* and there is also a short, soft *hweep* between bursts of song.

Nests are built just above the ground, in low shrubs and vegetation, and occasionally in low tree branches, ivy and creepers. Five or six eggs are laid; two broods are reared in the south of the range, while northern birds (which may not begin breeding until late May) raise only one brood.

Chiffchaffs arrive in Britain from early March and breed everywhere except for much of northern Scotland and upland areas of England and Wales. They are likewise absent from parts of Europe and northern Scandinavia. Different races of the species exist: the race that breeds in Spain has a different song, and the Scandinavian and Siberian races each have a different call. In Britain, the return migration begins in early August and continues to October. Many birds spend the winter in the Mediterranean region or farther south in West Africa. However, some stay in southern Britain for the winter, and Scandinavian birds may overwinter here. These birds feed on midges, caterpillars, flies, moths, aphids and spiders. The last two are probably the main food sources during colder months.

A slim, active bird, the chiffchaff looks similar to the willow warbler but has less interesting features. It is dull olive-brown above and buff below, with a pale stripe over the eye, and it has darker legs. Its distinctive chiffchaff *song is one of the first to be heard in spring. A small number of these birds spend the winter in southern Britain and will visit gardens, flitting through trees and bushes, hovering only to pick off caterpillars and flies.*

FACTS AND FEATURES

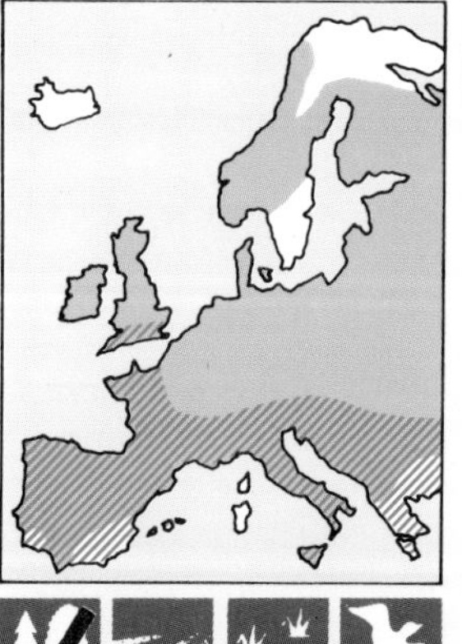

chiff-chaff-chiff-chiff-chaff-chiff-chaff...

Song A repetition of two notes, *chiff-chaff*, in irregular sequence.
Behaviour Sings from trees, often while moving.
Habitat Mixed woodland and shrubs.
Nest Of leaves, moss and stems in low bushes or brambles. Five or six eggs, one or two broods.
Food Flies, caterpillars, aphids and spiders.

WILLOW WARBLER *Phylloscopus trochilus*

A bright and active leaf warbler, the willow warbler has a greenish-brown back, yellowish-white underparts, a thin pale stripe over the eye and pale-coloured legs. Young birds have marked yellow underparts in the autumn. It is a summer visitor, normally arriving after the chiffchaff, its cascading, warbling song filling the woodland.

FACTS AND FEATURES

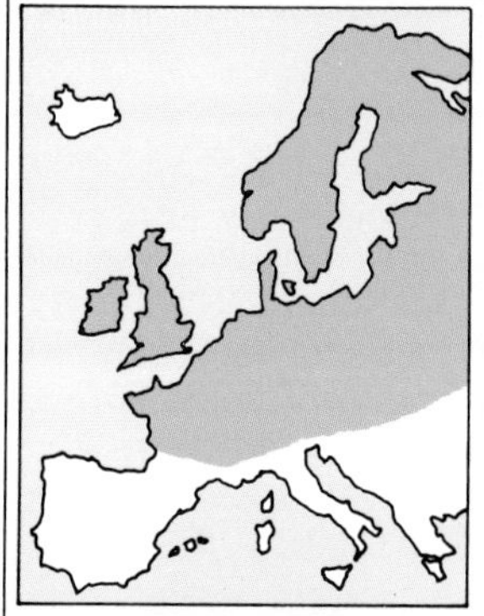

tswee-swee-swee-swee-swee-
swee-swee-sweeweeweeweetew

Song A slowly descending, rippling series of notes, rising slightly to end with a flourish.
Behaviour Sings from trees and bushes, often while feeding.
Habitat Deciduous and coniferous woodland, bushes, parkland and gardens.
Nest Of moss, stems, grass and bracken on the ground among vegetation. Six or seven eggs, one or two broods.
Food Flies, caterpillars, aphids, spiders and berries.

THIS ATTRACTIVE LITTLE BIRD is the commonest warbler in Britain and northern Europe, even ranging north of the Arctic Circle. It frequents deciduous and coniferous woodland, scrub, bushes, moorland, hedges, parks and gardens. Brighter than the dumpier chiffchaff, it has olive-brown upperparts, yellowish-white underparts – deeper yellow in autumn – and a fairly clear, yellow-buff supercilium ('eyebrow'). The legs are generally pale.

Like all leaf warblers, the willow warbler feeds actively, diligently searching out its insect food and often drooping its wings (but not flicking them like the chiffchaff). It feeds on small flies, weevils, moths, aphids, caterpillars and spiders. In autumn, elderberries and currants may be eaten.

The distinctive song is a cascading warble of liquid notes which gently descend the scale and end in a short, rising flourish. Song begins as birds arrive in early April and continues until early July, occasionally resuming in August and September. The male sings from trees and scrub, often perched in the open near the end of a branch, and frequently moving from one branch to another or feeding as he sings. The call is a plaintive *hoo-eet*, similar to a chiffchaff's, but tending to be more broken into two syllables. A quiet *chirrup* is sometimes given between songs.

The nest is built on the ground, usually in a hollow in a tuft of grass or among leaf litter. Six or seven eggs are laid and one brood, occasionally two, is reared.

The willow warbler does not breed in southern Europe, occurring there only on migration. There are two main races: a common, yellower race in western and central Europe, and a brown-and-white race in northern Europe. In spring, birds start to arrive in Europe from mid-March, reaching Britain in early April and northern Scandinavia in late May. They depart in August and September, with British birds spending the winter in tropical West Africa. Singing begins early in the year on the wintering grounds and some migrating birds sing while on their way north for summer. Willow warblers can be very aggressive when both breeding and migrating, sometimes attacking birds many times their size.

GOLDCREST *Regulus regulus*

EUROPE'S SMALLEST BIRD, the goldcrest – or golden-crested wren as it was once called – is an inhabitant of conifer woodland, oak woods, thickets and gardens. It likes spruce, fir, pine and cedar, but also takes to yews and exotic conifers in gardens. Goldcrests are small, plump birds, olive-green above and dull white below, with a double white wing-bar. The crest, from which they get their name, is a black-bordered yellow stripe down the centre of the crown. The male has an orange centre to the stripe, while the female's is all yellow. Noisy and quarrelsome, disputing males raise their crown feathers to display their crests. These warblers are very active, moving rapidly through the branches of conifers, often hovering to pick off food items.

The very high-pitched song is a series of double notes ending in a short flourish, *cedar-cedar-cedar-cedar-cedar-diddlydee*. It is given while the bird is feeding and can be heard virtually throughout the year, although less often in autumn and early winter. The call note is a high, thin *zee* often given in groups of three or four: *zee-zee-zee*.

Nests are built in conifers, suspended under a high branch near the top. Between seven and 10 eggs are laid and two broods raised. Sometimes the male takes care of the first brood while the female incubates the second clutch.

Goldcrests are resident in Britain, although some may move south and west in cold winters. They are found over most of Europe, but are absent from most of Spain and Portugal. Northern European birds migrate south; many come to Britain in autumn, large numbers arriving overnight along the east coast in October, and often coinciding with the arrival of woodcock. This has earned the goldcrest a local name, 'woodcock-pilot', due to the belief that such a small bird must have crossed the North Sea on the larger bird's back. A return migration occurs in March and April.

The diet consists of flies, aphids, weevils and spiders, as well as bugs, moths and other insects, and in winter they often join flocks of foraging tits. They frequently visit gardens during cold months to pick aphids from rose bushes or search ornamental conifers for food, and a few may take fat and bread from a bird table. Being chiefly insectivorous, goldcrests suffer badly in severe winters, although their large broods mean they can recover their numbers fairly quickly.

This tiny little bird – Europe's smallest – has a green back and a white wing bar, with pale underparts. Its bright crown stripe can be difficult to see and is orange-centred on the male and all yellow on the female. Young birds lack the crown markings.

FACTS AND FEATURES

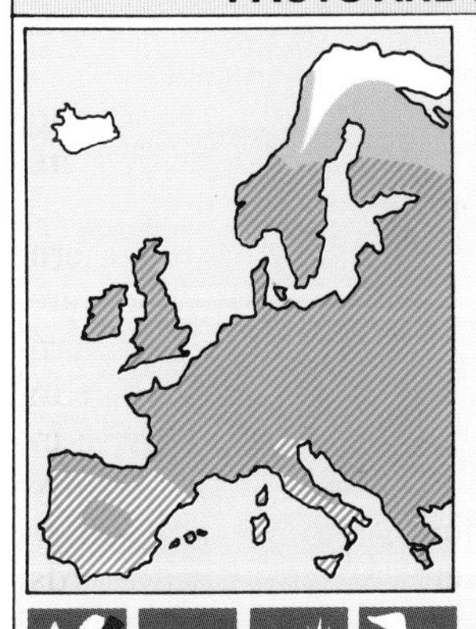

cedar-cedar-cedar-cedar-
cedar-diddlydee

Song An extremely high-pitched double note repeated rapidly and ending with a flourish.
Behaviour Sings from high in trees, often while moving.
Habitat Coniferous and mixed woodland, thickets and gardens.
Nest Of moss, lichen and spiders' webs, suspended under the branch of a conifer near its top. Seven to 10 eggs, two broods.
Fcod Flies, aphids, weevils and other beetles, moths and spiders.

FIRECREST *Regulus ignicapillus*

ONCE CALLED THE FIRE-CROWNED KINGLET, the firecrest certainly has a regal appearance. Similar in size to a goldcrest, it is immediately distinguished by its white supercilium ('eyebrow') bordered by a black crown stripe and black eye-stripe. The male has an orange-red strip down the centre of the crown, the female a yellow one. The general plumage is brighter than a goldcrest's, with whiter underparts, greener upperparts and a bronze patch on the side of the neck.

Firecrests flit through the foliage, looking for insects as they flutter and hover by leaves and twigs. They are more active than goldcrests, spending less time methodically searching for food, and they also take larger food items such as flies, caterpillars, moths and spiders. They also associate with deciduous woodland more than goldcrests: oak, beech and alder, as well as spruce, pine and fir, serve as homes. Out of the breeding season they are sometimes found in gardens, scrub and bushes. They can be very inquisitive birds and may come close to investigate a person, making shushing and squeaking sounds.

The song has similarities to a goldcrest's, but its high pitch identifies it as belonging to the 'kinglet'. Each phrase is a series of single high *zit* notes which become louder and faster at the end: *zit-zit-zit-zit-zit-zit-zit-zirtzirtzirt*. The song may be given while moving about and feeding; it can be heard virtually all year, but particularly from February to July, with the greatest output just after dawn in May and June. The call is a high but harsh *zit-zit-zit*.

Nesting is usually in conifers, occasionally juniper, or sometimes in ivy, and the nest is suspended from a branch. In Britain, eggs are laid in late May; in Spain, the first clutch is a month earlier. On average, from seven to nine eggs are laid, with up to 12 in the north of the range and only five in the south.

Firecrests have a more southerly distribution than goldcrests, being absent from Scandinavia but present in Spain and Portugal. British breeding birds appear to be summer visitors, arriving in May, but southern European birds are resident. In winter, this species can be seen in small numbers along the English coast from Merseyside right around to Kent. These wintering birds are of continental origin, probably from farther north and east.

FACTS AND FEATURES

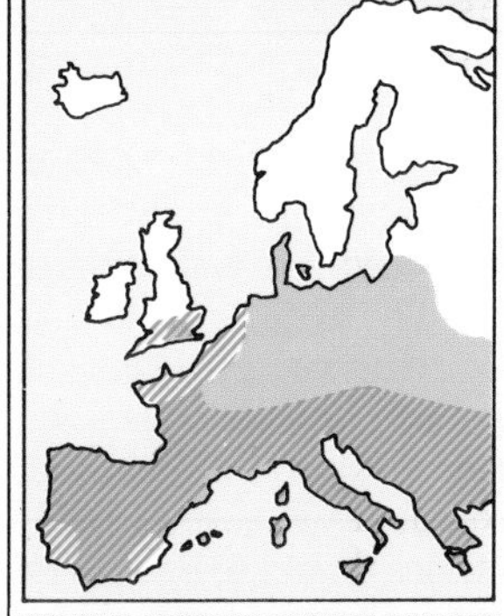

zit-zit-zit-zit-zit-zit-zit-zit-zit-zit-

zirtzirtzirt

Song A rapidly repeated series of high-pitched single notes.
Behaviour Sings from high up, and is especially vocal just after dawn in spring.
Habitat Deciduous and coniferous woodland.
Nest Of moss, hair and spiders' webs, suspended under a conifer branch, sometimes in ivy. Seven to nine eggs, two broods.
Food Flies, caterpillars, moths and spiders.

A brighter-looking relative of the goldcrest, the firecrest's white stripe above the eye (bordered by a black crown stripe and black eye-stripe) is a clear distinguishing feature. A close view also reveals a coppery tinge to the sides of the breast. A very active bird, it is seen less often in gardens than the goldcrest. It nests mainly in conifers, the nest suspended from a branch.

SPOTTED FLYCATCHER *Muscicapa striata*

A LOUD SNAPPING SOUND in a woodland clearing, and a small pale brown bird flying to a perch on a branch, are often the first signs of the spotted flycatcher. A bird which lives up to its name, it has a grey-brown back and white underparts with spotty brown streaks on its breast and head. Its method of feeding, in common with other members of its family, is to fly out from a perch and snatch a passing insect with an audible snap of the bill, before returning to the perch. If it fails to catch the insect first time it may give chase and try again, twisting and turning in its attempts. It can sit inconspicuously among foliage on the outer branch of a tree, or choose a more prominent perch on a bare branch, wire or post. Sometimes this bird perches on the ground and hops to pick up an insect. When perched it may flick its wings and tail occasionally. Woodland edges, parks, churchyards, gardens and the sides of wooded streams or lakes are its favourite habitat – wherever there are abundant flying insects.

For a bird with little colour, the spotted flycatcher has a surprisingly quiet and simple song. It is an unhurried series of thin, high notes – *sip, sip, see, sitti, see, see* – delivered from a perch in a tree. The short song period extends from early May, when the species arrives, until late June. The call note is a thin *tzee* and an alarmed bird gives an insistent *tzee-tzucc* or *tzee-tzucc-tzucc*.

The nest may be built on a tree or stump, or against a garden wall in creepers or ivy. Open-fronted nestboxes are used, as are old nests of swallows on shed beams. Four or five eggs are laid with one or two broods.

Spotted flycatchers spend the winter in tropical and southern Africa. They are one of the latest migrants to reach Britain, some arriving in the last half of April, but the majority not until mid-May. This ensures plenty of insects are on the wing, and breeding can commence rapidly. These birds are found throughout Europe with the exception of Iceland. They return south from late July but can still be present in October. The diet is mainly flies, butterflies and bees with some beetles, ants, dragonflies and damselflies.

FACTS AND FEATURES

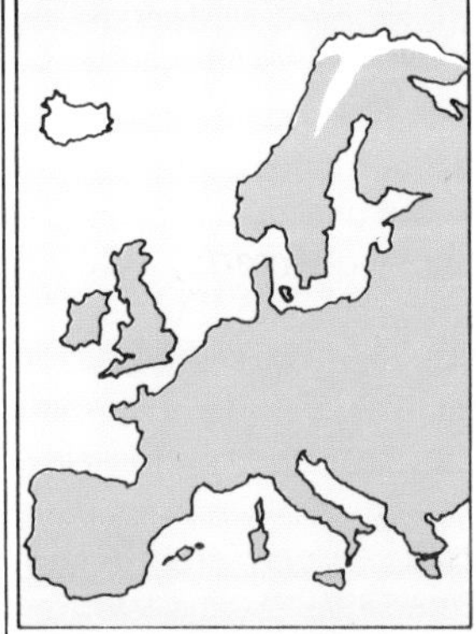

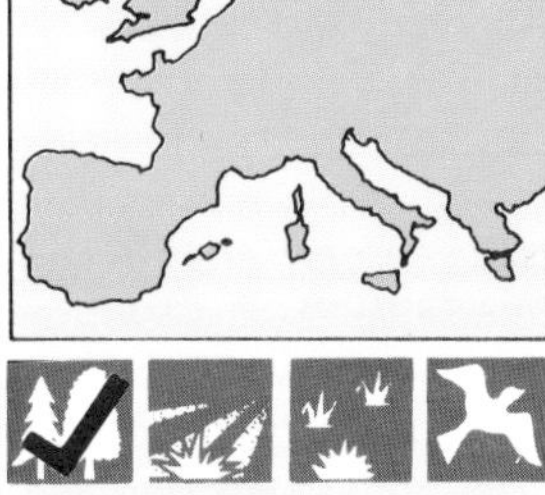

sip, sip, see, sitti, see, see

Song A quiet, simple song of squeaky notes with a pause between each.
Behaviour Sings while perched on a branch or fence.
Habitat Gardens, woods and parkland, and near water.
Nest Of moss, wool and hair in a tree hole, in creepers on a wall and often in an open-fronted nestbox. Four or five eggs, one or two broods.
Food Flies, butterflies, bees, ants, beetles and dragonflies.

The spotted flycatcher is a distinctive bird even though its plumage – greyish-brown above and white spotted with brown below – is not very bright. However, its fly-catching action, darting out from a perch and catching an insect with a loud snap of its beak, makes its presence obvious. The spotted flycatcher is usually easy to pick out as it chooses a prominent perch, where its upright stance and pale breast stand out. It will often nest in creepers on a garden wall.

PIED FLYCATCHER *Ficedula hypoleuca*

THE UNMISTAKABLE black-and-white plumage of the male pied flycatcher is surprisingly good camouflage when this bird is perched in the dappled sunlight of an oak wood. Areas of sessile oaks in western Britain form the typical habitat, as well as alder and birch along rivers and streams. Generally this species prefers open deciduous woodland in upland valleys and foothills, and also parks and gardens with mature trees. Mixed woodland is used in northern Europe.

Both males and females have white underparts with a conspicuous white wing patch and white outer tail feathers. Upperparts are black in the male and brown in the female. An active bird, the pied flycatcher takes insects like its spotted namesake; it may drop to the ground, or hover to pick a caterpillar from a leaf, or cling to the trunk for a grub hidden in the bark. The diet includes flies, beetles, moths and ants as well as caterpillars and grubs. More nervous-looking than the spotted flycatcher, it indulges in much wing-flicking and occasionally cocks its tail when excited.

The song is a repeated double note, *zee-it, zee-it, zee-it*, followed by a jangling trill or warble. The singing bird is usually perched well up in a tree among the leaves, but it may use a telegraph wire or exposed rock. Singing commences in late April, continuing until the eggs are laid, and is heard infrequently from the end of May until late June. Call notes are a sharp *whit* and a short *tic*, or a combination of the two.

The nest is built in a tree hole, which explains the preference for mature oak. Hole-fronted nestboxes are readily occupied in less mature woodland. Indeed, these boxes have enabled populations to increase and spread in some western areas of Britain, where they have been used on reserves. From five to eight eggs are laid and one brood is reared.

European pied flycatchers spend the winter in the savanna forests of tropical Africa. They arrive in most parts of Europe from mid-April to mid-May, with those in the north not arriving until June. The return migration is from August until October. Many northern European birds may occur on the east coast of Britain in late August and early September, if weather conditions force them across the North Sea.

FACTS AND FEATURES

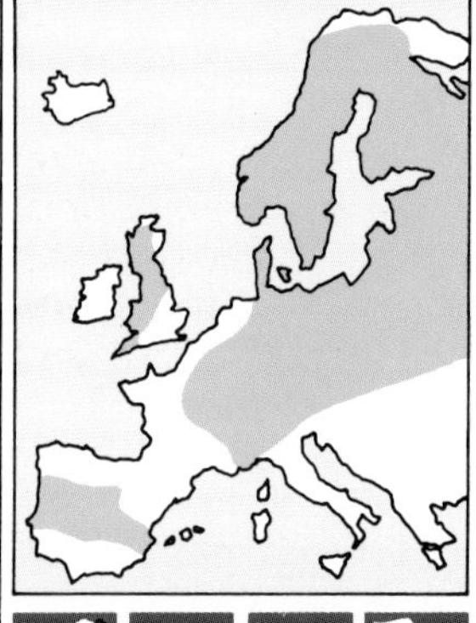

Song Similar to a redstart's, usually beginning with repeated *zee-it* followed by a jangling trill.
Behaviour Song given from a high perch on a tree.
Habitat Mature woodland, especially oak and birch, and parkland.
Nest Of moss, leaves and roots in a tree hole or open-fronted nestbox. Five to eight eggs, one brood.
Food Flies, beetles, caterpillars and ants.

The contrasting black-and-white plumage of the male pied flycatcher is surprisingly difficult to see when the bird is perched in dappled foliage in a tree. Only its fly-catching activities give it away. The upperparts of the male are black; in the female they are brown, but both have a conspicuous white wing patch and white outer tail feathers. The pied flycatcher favours mature deciduous woodland and readily uses hole-fronted nestboxes.

MARSH TIT *Parus palustris*

Difficult to distinguish from its close relative, the willow tit. Its glossy black cap, tidy bib and lack of a pale wing panel are the best distinguishing clues to its identity if it is silent. Its loud calls and distinctive song are also a guide. It will visit gardens, especially in the winter, and will occasionally come to bird tables for nuts or fat.

FACTS AND FEATURES

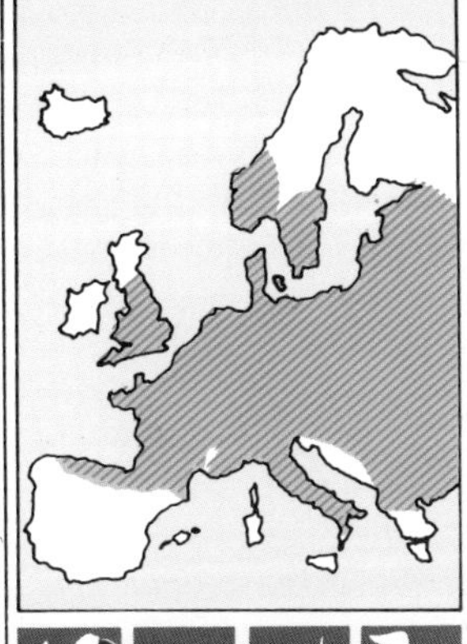

chip-chip-chip-chip-chip,
pitchaweeoo

Song A loud repeated *chip*, sometimes with a variation of the call, *pitchaweeoo*.
Behaviour Sings from medium height in trees and bushes.
Habitat Deciduous woodland and copses, with no preference for marshy areas (despite the name).
Nest Of moss and hair in tree holes and rarely hole-fronted nestboxes. Seven or eight eggs, one brood.
Food Caterpillars, beetles, aphids; seeds from thistle, rowan, yew and honeysuckle; beechmast and sunflower seeds. Visits tit feeders in cold weather.

DESPITE ITS NAME, the marsh tit does not inhabit marshland. It prefers dry or damp deciduous woodland, with beech and oak and a ground cover of shrubs. It may also be found in willow and alder along rivers, and has a liking for yew trees. In winter it visits hedgerows and gardens.

The marsh tit looks very similar to the willow tit, but always appears cleaner and brighter. The upperparts are brown, the black cap glossy, the black bib small and neat, and the cheeks and underparts white. Like all members of the tit family it is an active bird, although it spends less time in the tree canopy than its cousins and tends to be found in shrubs and small trees, nearer to the ground. Marsh tits are usually seen in pairs or family parties and forage with other tits in winter.

The song is a simple series of repeated notes, *chip-chip-chip-chip*..., or sometimes a two-syllable *chippi-chippi-chippi*.... Occasionally there is another type of song, using a ringing *pitchaweeoo*. Singing is usually from a tree or bush and the main period is from mid-January to mid-May. The call of the marsh tit is very distinctive, although it is sometimes confused with one of the many calls given by great tits – a loud, sharp *pitchoo*, often followed by a series of nasal notes, *tchaa-tchaa-tchaa*.

The nest is built in a hole fairly close to the ground, often in a hollow branch or a crack in a stump. Holes in walls and even in the ground may be occupied if there are no suitable trees, but these birds rarely take to nestboxes. They compete with blue and coal tits for nest sites, giving way to great tits and nuthatches. Seven or eight eggs are laid and only one brood is produced.

Marsh tits are found over most of Europe except for the far north and the Iberian Peninsula. They are absent across Asia, but in China and Japan there is a separate population. These resident birds stay paired through the winter, remaining on their territory; young, unpaired birds roam with mixed tit flocks. They feed low down in woodland, taking weevils, grubs and other insects, and selecting larger prey items than blue or coal tits. In autumn their diet switches to nuts, seeds and berries, in particular beechmast (beech nuts), and rowan and honeysuckle berries. Sunflower seeds are a favourite at the bird table. Marsh tits are food hoarders and hide nuts and seeds in moss, retrieving them often several days later.

WILLOW TIT *Parus montanus*

IN BRITAIN the willow tit was not distinguished as a separate species from the marsh tit until 1897. At a distance these two species look identical. However, closer scrutiny reveals that, compared with a marsh tit, the cap of a willow tit is not glossy but a dull sooty-black, the bib is slightly larger and not so neat, it has buffish flanks, and the pale edgings to the wing feathers form a noticeable panel on each wing. Because these features are not always visible, the best way of distinguishing the two is by voice. Willow tits prefer damp alder and birch woods, also elder and willow thickets, and are even found in coniferous trees – in northern Europe they inhabit birch and spruce. Like marsh tits, they spend most of their time in shrubby growth near the ground rather than in the tree canopy.

Two types of song can be discerned, but neither is used with any frequency. The commoner type is a clear *piu-piu-piu*, the other a quiet warbling phrase. Calls are heard much more often and are the main aid in identification. A low, harsh *tchair-tchair-tchair* is often preceded by a high *zi-zi*, or a *chichit*, which can be used as a separate call.

The nesting site is a hole in a rotten trunk or stump, but, unlike the marsh tit, the willow tit excavates its own hole. Birch, willow, elder and alder are favourite trees. This bird does not normally use a nestbox, but has been attracted to one placed on a stump and filled with wood chippings which it can 'excavate'. Eight or nine eggs are laid and one brood raised.

Willow tits are residents and do not move from their breeding territories. They have a more northern distribution than marsh tits and those in Scandinavia are larger and paler. These northern birds may move in winter if their food supplies disappear, being rare vagrants to Britain. Both willow and marsh tits are absent from Ireland, and in Scotland only the willow tit breeds, and then only in the south. Willow tits feed on similar food items to marsh tits, but do not forage on the ground much, and with their finer bills they do not take beechmast (beech nuts). They are also more likely to visit garden bird tables for peanuts and fat.

In northern Scandinavia there is another species of tit, the Siberian tit, *Parus cinctus*, like a large brown-capped willow tit with a hoarse *si-chee-chee-chee* call.

Its dull, sooty black cap, larger and more untidy bib and pale wing panels help to distinguish the willow tit from the marsh tit and, if heard, it can easily be identified by its markedly different call. It is one of the few tits to excavate its own nest hole, for which it needs rotten wood. The female makes the hole and the nest is made out of wood chips, plant fibres and hair. The willow tit is more likely than the marsh tit to be found in damp woodland.

FACTS AND FEATURES

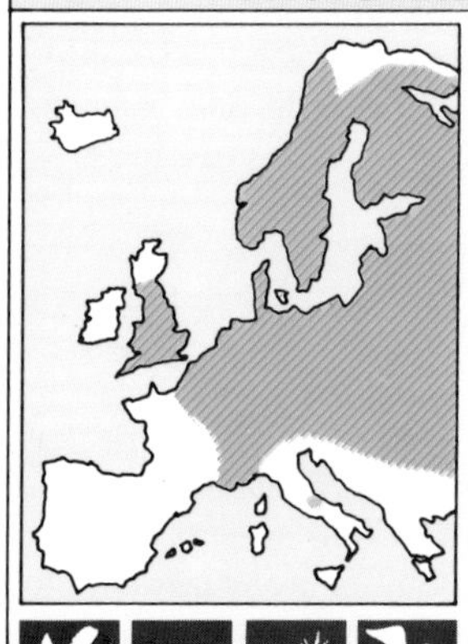

piu, piu, piu

Song A clear, repeated *piu-piu-piu*, and a quieter warble.
Behaviour Not very vocal, singing from low in trees, bushes and tall undergrowth.
Habitat Damp woodland and copses, both deciduous and coniferous.
Nest Of moss, hair and wood fibre in a tree hole excavated in rotten wood. Nestbox may be used if filled with wood chippings. Eight or nine eggs, one brood.
Food Caterpillars and other insects, some seeds. Visits feeders for peanuts.

CRESTED TIT *Parus cristatus*

THIS BIRD OF CONIFEROUS WOODLAND is similar to the marsh and willow tit, but is immediately identified by its charming black-and-white speckled crest. It spends its time high in the upper branches of pine trees, exploring the buds for insects. In common with other tits it is very agile, hanging upside down on a branch or searching up bark like a treecreeper. In Britain, it is mainly confined to mature Scots pine, but has taken advantage of some new conifer plantations. It can be seen feeding in birch or alder where these trees are mixed with conifers, and in winter it may range out of its normal habitat, sometimes into gardens.

There is no real song – a louder version of the call is the nearest this tit comes to singing. The call is a distinctive churring trill, *chirrr, chirrr, chirrr-r-r*, sometimes with a high *zi-zi-zi* before it, and also given as a double note, *zi-chirrr*.

Like willow tits, crested tits excavate their nest holes in rotton trunks and stumps, mainly in pine but occasionally in other trees. Nestboxes may be used in a suitable habitat if filled with wood chippings to remove. Five or six eggs are laid and one brood reared.

In Britain, crested tits are found chiefly in small, isolated regions of Scotland where areas of the old Caledonian pine forests remain. On the Continent they are absent only from northern Scandinavia and parts of southern Europe. Scottish birds are darker than continental ones, and northern and central European birds also differ from eath other. These tits are residents, rarely moving far from their breeding territories, although some birds do wander, since both continental races have been seen in southern England. They search for insect food in high branches, but also investigate lower branches and tree trunks as well as feeding on the ground. They eat caterpillars, aphids, spiders, seeds from pine cones and berries; they also come to bird tables near their territories in winter.

This member of the tit family is made all the more attractive by its black-and-white speckled crest. It lives in coniferous forests and is mainly confined in Britain to the old Caledonian pine forests of Scotland. Like the willow tit, it is the female who excavates the nest hole in a stump of rotting wood. It will venture into gardens near its breeding area.

FACTS AND FEATURES

chirrr-chirrr-chirrr-r-r

Song A repeated, trilling *chirrr*.
Behaviour Sings from trees, often near the top.
Habitat Mature pine forest in Britain but found in mixed woodland elsewhere in Europe.
Nest Of moss and hair in a tree hole. Occasionally uses nestboxes. Five or six eggs, one brood.
Food Caterpillars, aphids and spiders, some seeds and berries.

COAL TIT *Parus ater*

FACTS AND FEATURES

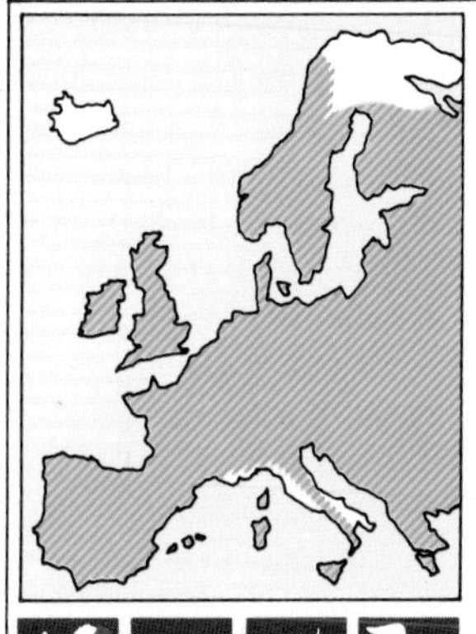

see-too, see-too, see-too...

Song A piping *see-too* repeated like a great tit.
Behaviour Sings from trees, usually from a high perch.
Habitat Mixed woodland, parks and gardens, generally with conifers.
Nest Of moss and hair in tree hole or hole in a bank. Readily uses hole-fronted nestbox. Seven to 11 eggs, one brood (two in central Europe).
Food Flies, caterpillars, weevils and other beetles, spiders, seeds and nuts. Takes peanuts, seeds and fat from bird feeders.

THE SMALLEST OF THE TITS, with a character all its own, the coal tit can appear timid – at a bird table dispute it usually gives way to blue and great tits. This dainty bird has grey upperparts, a double white wing bar, a black cap with a white patch at the back of the neck, and a large black bib; the underparts are buff-white. It is one of the few species that has derived benefit from the planting of large areas of conifers, since it inhabits coniferous and mixed woodlands, suburban gardens with conifers and, in southern Europe, oak and beech woods.

The coal tit has a narrower bill than other tits, which enables it to pick out insects between conifer needles and seeds from cones. It feeds on weevils, grubs, flies, caterpillars and spiders in summer, and turns to conifer seeds, beechmast and other nuts in winter. Coal tits readily visit bird tables in winter and are particularly fond of peanuts, which they remove and store in a nearby conifer or among moss on the ground. But great tits often take advantage of this behaviour and steal the nuts from their hiding places!

The song is a piping *see-too, see-too, see-too*... or *teechuee, teechuee, teechuee*. Singing is often from high in a conifer and carries for some distance. Song period is from January to late July, and to a lesser extent in autumn and early winter. The commonest call is a plaintive *tseu* and also a high *tsee-tsee-tsee* similar to other tits.

Nests are built in tree holes, holes in walls, among tree roots and even in holes in the ground. Most nests are fairly low down. The bird may use a nestbox with a small entrance hole, provided it does not have to compete for it with blue or great tits. There are from seven to 11 eggs, and usually only one brood.

Coal tits are found from Europe right across central Asia to Japan. They are mainly resident in Europe and stay very close to their home territories throughout the year. They join the winter flocks of various tits that forage around woodlands and gardens. In Scandinavia some birds may move south, and in parts of southern Europe birds are found in the winter only. Continental coal tits have been involved in large 'irruptions', with considerable numbers of birds moving south and west from northern and central Europe, probably due to high populations and food shortages.

A common visitor to bird tables, particularly if there are coniferous trees nearby, the little coal tit has a grey back, double white wing bars, black cap with a white patch on the nape of its neck and black bib making it easily identifiable. A timid small bird, it is usually dominated by the more aggressive blue and great tits at the bird table, where, like them, it feeds on nuts or fat.

BLUE TIT *Parus caeruleus*

ONE OF THE MOST COLOURFUL and entertaining of garden birds, the blue tit is found from woodland to city centre. It has yellow underparts, a green back, blue wings and crown, and white cheeks. It is lively, cheeky, acrobatic and noisy, and can be attracted by virtually anyone who puts out peanuts at a bird table. These tits are found in deciduous woodland, hedgerows, parks and gardens. Oak, beech, birch and ash are the most popular trees, providing both nest holes and food.

The blue tit's song is as cheerful as its nature: a series of high notes merging into a trill, *tsee-tsee-tsee-tsuhuhuhuhuhu*.... Singing is from a tree or bush, mainly from January to June. The commonest calls are a high *tsee-tsee-tsit* and an insistent *tsi-tsi-tu-tu-tu*.

The typical nest site is a hole or crack in a tree, or a hole-fronted nestbox where there is no natural site. In the absence of both, blue tits may use holes in walls, pipes, letter boxes, flowerpots, street lamps – anything with a suitable entrance hole. They lay between seven and 14 eggs and have one brood in Britain. In southern Europe, two smaller broods are regularly reared.

Blue tits are resident across Europe; only birds from Scandinavia regularly move south. Continental blue tits 'irrupt' on occasion and come to Britain, being distinguished by their brighter colours. These tits feed on insects in summer, in particular the caterpillars of winter moths and other moths, also weevils, grubs, aphids and spiders. The small bill means that in winter this tit can pick out small hibernating insects and also take beechmast from the tree, but it is unable to get at insects in conifers.

A welcome visitor to bird tables, the blue tit has blue wings and crown, a greenish back, yellow underparts and white cheeks. It nests in a variety of holes and is always ready to inspect a nestbox. Its agility at hanging from branches in search of food is entertaining, as are its antics on hanging feeders or at the bird table. Blue tits may be less welcome when they remove putty from windows or peck open the tops of milk bottles.

Blue tits are famous – or infamous – at finding new food sources. An individual near Southampton was credited with the first opening of a milk bottle top to get at the cream beneath. They are inquisitive and have been known to enter houses in search of food. They have also been labelled 'hooligans' because of certain habits, such as removing putty from around windows and 'paper-tearing' with birds actually stripping wallpaper off walls. One blue tit pecked out all the red flowers from a floral cushion cover! In some years when this behaviour has been prevalent, there have been unusually large irruptions of blue tits from the Continent, so it may be connected with food-searching instincts. At bird tables, blue tits eat peanuts, coconut, fat, seeds and cheese, taking food from suspended feeders that few other birds can reach.

FACTS AND FEATURES

tsee-tsee-tsee-tsuhuhuhuhuhu

Song A repeated high *tsee* which merges into a trill.
Behaviour Sings from varying heights in trees.
Habitat Deciduous woodland, copses, parks and gardens.
Nest Of moss, grass and hair in a tree hole or hole in a wall. Readily uses hole-fronted nestbox. Seven to 14 eggs, one brood (two in central Europe).
Food Caterpillars, grubs, aphids, spiders, fruit buds, fruit, beechmast and seeds. Commonly visits feeders for peanuts, fat and seeds.

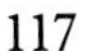

GREAT TIT *Parus major*

THIS IS THE LARGEST of the European tits, and perhaps the noisiest. It is a familiar bird with its green back, blue wings with a white wing bar, yellow underparts with a black stripe down the centre, and black head with white cheeks and a pale nape patch. It inhabits woodland, hedges, parks, gardens and almost anywhere else with deciduous trees. Oak, ash, beech and birch, and also yew (but few other conifers), are favourite trees. The great tit's size makes it the most dominant of the tit family, and it is aggressive not only to other tits but even to bigger birds that compete for food at the bird table.

The song is a loud, ringing *tea-cher, tea-cher, tea-cher*, sometimes likened to a squeaky bicycle pump. It is given from trees and bushes almost all year round, but mainly from January to June. The calls are extremely varied and one bird may make more than 40 different sounds. A chaffinch-like *chink*, a high *tsee* and a nasal *cha-cha-cha* are some of the commoner calls.

The great tit nests naturally in tree holes, but like the blue tit it will use many other, unnatural sites. Nestboxes are very popular, provided the hole is big enough. From five to 11 eggs are normally laid and one brood is reared, but in southern Europe there is a smaller clutch and two broods.

Great tits are found across Asia in a variety of plumages, Japanese birds having white underparts. They are resident in Europe, except for northern Scandinavia, where they are present in summer. They may move from high ground in winter – Scottish birds, for example, descending into valleys and glens where feeding is easier. As with blue tits, there are occasional arrivals of continental birds in Britain when food supplies are short. Great tits spend more time feeding on the ground than other tits, and with their larger bills they are able to consume bigger seeds and nuts in winter. However, their size precludes them from the small twigs and leaves at the tips of branches, which blue and coal tits reach. They feed on caterpillars (mainly winter moth, mottled umber and cabbage white), bees, aphids and spiders. It is estimated that a pair of great tits with young can destroy up to 8,000 caterpillars in three weeks. Winter foods are beech-mast, hazelnuts and yew seeds. This tit also visits bird feeders for peanuts, sunflower seeds, fat and cheese.

Another large European tit, confined to the south-east, is the sombre tit, *Parus lugubris*. It resembles a large willow tit with a more extensive bib and darker brown back. This species has a distinctive disyllabic song and churring calls.

The largest of the tits that regularly visit gardens, the great tit will use nestboxes and will feed from hanging nut feeders. The male usually has a broad black stripe down the centre of his yellow breast whereas the female has a thinner, often broken, stripe. Its ringing call of tea-cher, tea-cher *is unmistakable.*

FACTS AND FEATURES

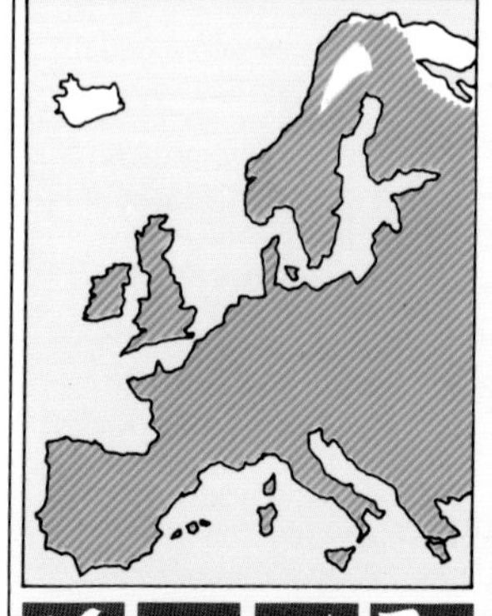

tea-cher, tea-cher, tea-cher

Song A loud, repeated *tea-cher*.
Behaviour Sings from tree canopy and bushes.
Habitat Deciduous and mixed woodland, copses and gardens.
Nest Of moss, grass and roots in a tree hole. Readily uses nestbox. Five to 11 eggs, one brood (two in southern Europe).
Food Caterpillars, aphids, beechmast, seeds and fruits. Takes peanuts, seeds and fat at feeders.

LONG-TAILED TIT *Aegithalos caudatus*

THE LONG-TAILED TIT is not a true tit, being anatomically different from members of the tit family. Its black-and-white plumage is distinctive and its small body, tiny bill and very long tail make it recognizable even in silhouette. It is a gregarious bird, travelling in parties which sometimes number several dozen. Such groups are noisy and active, staying together throughout the winter. In flight they undulate, their tails waving behind them. In trees they are acrobatic, hanging from twigs and fluttering between branches, hardly staying still for a moment. These birds are found in deciduous woodland, hedgerows, scrub, parks and suburban gardens.

The long-tailed tit does not have a clear song, but it may utter phrases which contain various call notes. One call is a sharp, purring *tsirrrrp* with an almost trilling quality; another is a short *tupp*; a third is a high *zee-zee-zee* repeated frequently, often in flight.

The nest is a most intricate construction of moss, gossamer and lichens, lined with feathers (more than 2,000 feathers have been counted in one nest). It is shaped like an oval ball with a hole near the top, and is concealed in a thick bush or shrub or in the fork of a tree. From eight to 12 eggs are laid and one brood is reared. It is not uncommon to find three birds tending a nest, the extra being a young unmated bird or one that has failed to raise its own offspring.

Long-tailed tits are found all over Europe except for northern Scandinavia. In northern and eastern regions there is a distinct race, occasionally seen in Britain, that has an all-white head. Birds are resident in Britain, staying near their breeding areas. On the Continent long-tailed tits may move some distances, and continental birds have reached Britain when large-scale invasions of tits have taken place. Their size makes them vulnerable in severe winters and they may roost communally on a branch, forming a tight ball of birds. When snow and freezing temperatures make food hard to find, large numbers can perish. The long-tailed tit feeds on caterpillars, ants, weevils, spiders and seeds.

FACTS AND FEATURES

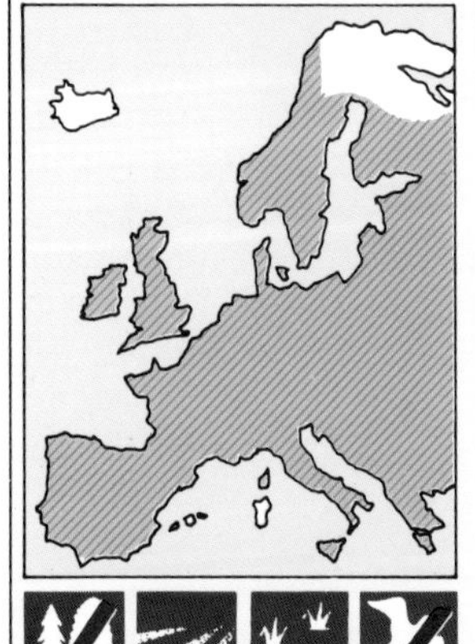

see-see-siu

Song A repetition of its *zee* call with other notes.
Behaviour Very active, sings and calls from trees and bushes.
Habitat Hedgerows, woodland edges, parks and gardens.
Nest Of moss, cobwebs, lichens and hair, oval in shape, with an entrance hole near the top. Usually placed in thick bush or shrubs. Eight to 12 eggs, one brood.
Food Caterpillars, weevils, ants, some spiders and seeds. Rarely visits feeders for peanuts and fat.

The very long tail and small body of the aptly named long-tailed tit make it very distinctive, as indeed is its striking black-and-white plumage. It is usually seen in flocks which can include up to several dozen in the winter. Its nest is a marvellous construction of spiders' webs and lichens, lined with feathers. In cold weather long-tailed tits can often be seen roosting close together on a branch.

NUTHATCH *Sitta europaea*

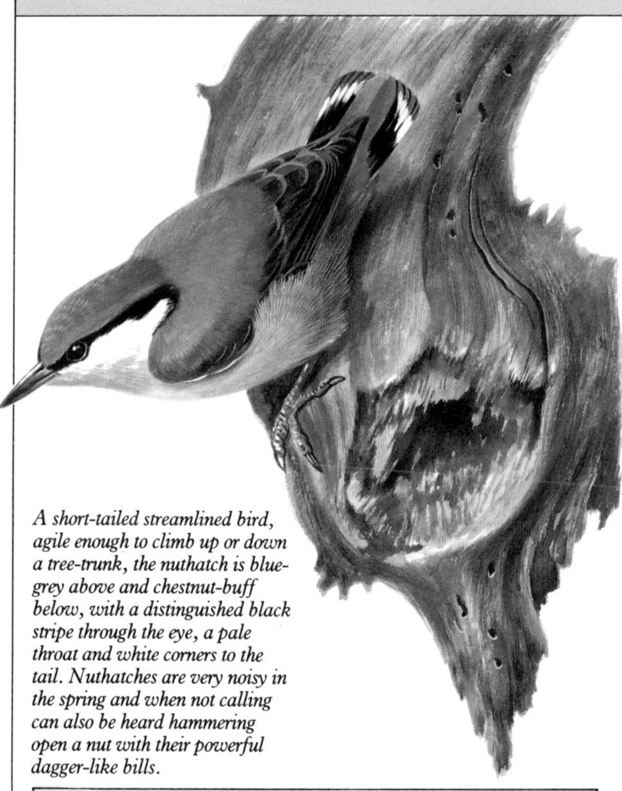

A short-tailed streamlined bird, agile enough to climb up or down a tree-trunk, the nuthatch is blue-grey above and chestnut-buff below, with a distinguished black stripe through the eye, a pale throat and white corners to the tail. Nuthatches are very noisy in the spring and when not calling can also be heard hammering open a nut with their powerful dagger-like bills.

THIS SLIM, SHORT-TAILED BIRD is unique in its ability to run both up and down the trunk of a tree. Unlike treecreepers and woodpeckers, it does not use its tail for balance when climbing, and moves jerkily along a vertical surface. It often stays high in the foliage and may only be noticed when it calls or hammers at a nut with its stout, dagger-like bill. Plumage is blue-grey above and chestnut-buff below, with a black stripe through the eye, a pale throat and white corners to the tail. The flight is undulating but maintains a steady height, not dropping to the base of a tree like a treecreeper. The nuthatch can also run swiftly along branches and readily hops on the ground in search of food. It inhabits mature deciduous or mixed woodland with oak, beech and chestnut. Parks, large gardens and hedgerows with tall mature trees are also suitable.

A noisy bird, the nuthatch has a variety of calls and songs. One is a ringing *qui-qui-qui-qui-qui*. . . , another a rapid, trilling *chichichichichi*. The song is usually delivered from the branch of a tree but sometimes from a prominent perch on an outer branch or at the tree's top. Singing begins in early January and continues to mid-May. The commonest call is a ringing *chwit-chwit, chwit-chwit-chwit* and also a high, tit-like *tsit-tsit-tsit*.

The nest is usually built in a hole in a tree branch or trunk, the entrance being plastered with mud to make it the right size. Nestboxes are used, also being given a mud-lined entrance, and mud may be used to fill any cracks in the box. The bird lays from six to 11 eggs and has one brood.

The nuthatch is resident in Europe and does not move far outside its territory, although it may forage with winter flocks of tits. Its sedentary nature means that there are many different races, perhaps the most distinctive being Scandinavian birds with their white underparts and chestnut on the flanks. Nuthatches feed on caterpillars, beetles, flies, grubs, earwigs and spiders. In winter the long, stout bill cracks open nuts such as hazelnuts and acorns, and this bird also eats beechmast (beech nuts) and yew seeds. At the bird table it especially likes peanuts and sunflower seeds, flying off to wedge them in a tree, and crack and eat them. It may store food, hiding it in tree crevices.

In southern Europe there is a similar species, the rock nuthatch, *Sitta neumayer*, which has white underparts and no white tail corners. Its song is more trilling and it sings, nests and feeds among rocks rather than trees.

FACTS AND FEATURES

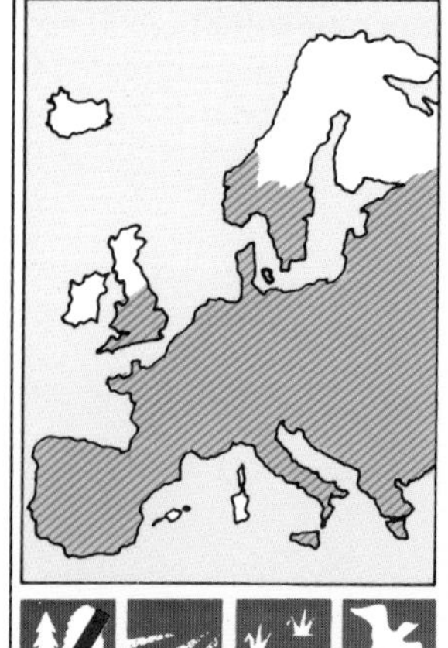

qui-qui-qui-qui-qui

Song A ringing *qui-qui-qui-qui-qui* and also a rapid trill.
Behaviour Sings from the tree canopy and also from an exposed branch.
Habitat Mature woodland with oak, parks and gardens.
Nest Tree hole lined with wood chippings. Mud plastered around entrance. Hole-fronted nestboxes are also used. Six to 11 eggs, one brood.
Food Caterpillars and other larvae, spiders, earwigs, hazel nuts, acorns, beechmast and other seeds. Readily visits feeders for peanuts and sunflower seeds.

TREECREEPER *Certhia familiaris*

THIS INCONSPICUOUS LITTLE BIRD has upperparts streaked with brown, and a white supercilium ('eyebrow') and underparts. As its name implies, it spends its time creeping along the trunks and branches of trees. It always climbs up and never goes down (unlike the nuthatch) since it uses its stiff-pointed tail for support, clinging close to the trunk. With jerky movements this bird seeks out small insects and grubs with its long, thin bill. It eats a variety of insects, their eggs and larvae, as well as spiders. Treecreepers do not come to bird tables, but have been known to take crushed nuts placed in tree crevices. A feeding bird gradually works its way up the tree, spiralling around the trunk and larger branches, then flying to the base of the next tree and starting again. Sometimes it feeds on smaller twigs, hanging upside down like a tit. This species inhabits mature deciduous and mixed woodland in Britain, and also mature conifers. On the Continent it is found mainly in upland coniferous woodland, being replaced by the short-toed treecreeper in deciduous woods.

The treecreeper's mouse-like behaviour would make it difficult to detect, but it has a loud call and sweet song. The latter is a delightful, high-pitched series of slightly descending notes ending with a flourish, *tsee-tsee-tsee-tsee-sissi-sissi-swee*. The bird sings while climbing or motionless on a branch, and occasionally in flight. The main song period is from February to May. The call is a loud *tseuu*, often repeated rapidly.

The nest is built in a crack in a tree or behind a flap of bark or ivy stems. This bird also uses a specially constructed wedge-shaped nestbox with a side entrance, placed on a tree trunk. It lays six eggs and occasionally rears a second brood.

Treecreepers are resident birds, and like tits and nuthatches, they do not move far outside their territories. An insectivorous diet makes them susceptible to cold winters and in severe conditions many perish. They roost behind bark and ivy and have also adapted to an introduced ornamental conifer, the *Wellingtonia*, which has a soft, flaky bark in which the bird excavates a roosting cavity.

FACTS AND FEATURES

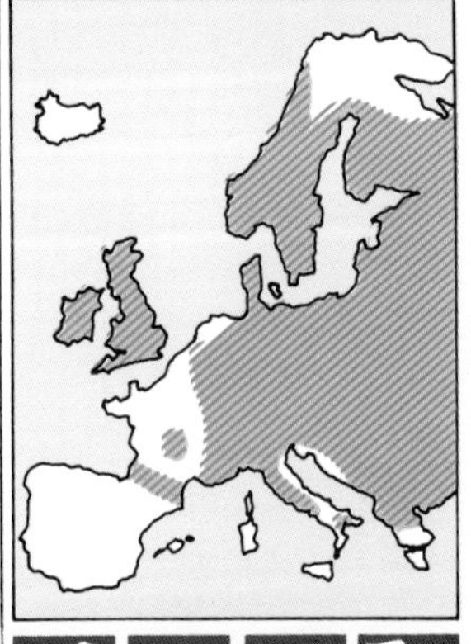

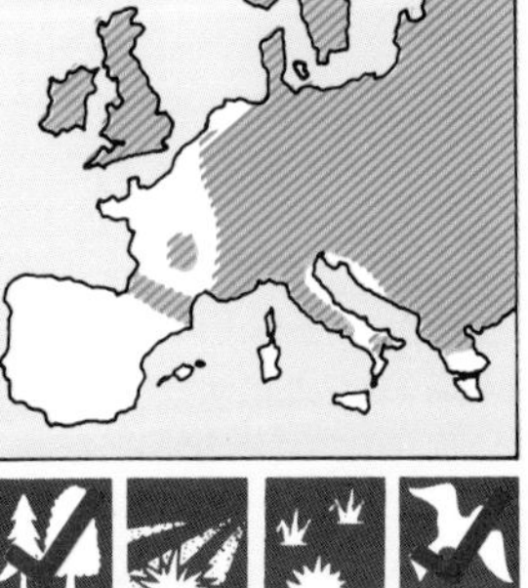

tsee-tsee-tsee-sissi-sissi-swee

Song High notes descending to a distinctive ending, *sissi-sissi-swee*.

Behaviour Sings while climbing up tree trunks and branches, occasionally in flight.

Habitat Mature conifers and deciduous woodland, parks and gardens.

Nest Of moss, grass, roots and twigs in a tree crevice, behind ivy or in a hole in rocks. Uses a wedge-shaped nestbox. Six eggs, one or two broods.

Food Insects, eggs and larvae, such as weevils and caterpillars, and spiders. Takes crushed nuts from tree crevices.

A well-camouflaged bird, the treecreeper spends its time clinging onto the bark of trees, where its brown-streaked upperparts make it hard to spot. It looks more like a small mouse as it climbs up a trunk with jerky movements, working up the tree before flying down to the base of the next one. Its fine, curved bill is ideal for extracting grubs and spiders from cracks in the bark. It has an attractive high-pitched song.

SHORT-TOED TREECREEPER *Certhia brachydactyla*

THIS SPECIES IS DIFFICULT to distinguish with certainty from the treecreeper, since although it has shorter claws, this feature cannot be used in the field. There are some plumage differences which may help, but song and calls are the best guide to separating these two very similar species.

Compared with the treecreeper, the short-toed treecreeper has greyer upperparts. It does not show the white streaking on the upperparts, but has contrasting dark streaks, giving a more broken pattern. The flanks are usually buff and the supercilium ('eyebrow') is shorter than a treecreeper's. A good view of the bill shows that the lower mandible and sides of the upper mandible are pale, while a treecreeper has an all-dark upper mandible, giving it a dark-tipped appearance. Habitat and distribution can also help to separate them, since treecreepers are found in conifers while short-toed treecreepers are mainly birds of deciduous woodland. They are also found in open woodland, coppices, parks and gardens and are called 'garden creepers' in France. Behaviour in the two species is identical, with the same meticulous searching of trunk and branches for food.

FACTS AND FEATURES

tsee-tsee-tsoo-tsoo-tsee-tsee-tsee

Song A high, rhythmic phrase of six or seven notes.
Behaviour Sings from a tree trunk and more often from an exposed branch.
Habitat Prefers deciduous woodland and some pines.
Nest Of moss, grass, roots and twigs behind bark or ivy. Six or seven eggs, one or two broods.
Food Caterpillars, weevils and spiders, some seeds.

The song is lower than that of a treecreeper and has totally different phrasing, being shorter and noticeably more forceful. It is made up of six or seven notes, *tsee-tsee-tsoo-tsoo-tsee-tsee-tsee*, with the third and fourth notes falling and the last three rising. It is sometimes written as *teet-teet-teet-e-roi-i-titt*. The singing bird is usually perched on a branch, but may be climbing a tree trunk. The call notes are varied, but a shrill, explosive *zeet* or *tseep* is characteristic.

The nest is built behind bark or ivy. Six or seven eggs are laid and occasionally two broods are reared. Wedge-shaped nestboxes may be used, as with the treecreeper.

Treecreepers are absent from Spain, Portugal and parts of France, and here only the short-toed species is found. In the Channel Islands, too, only short-toed treecreepers are present, living on wooded valley slopes. They are resident and do not move far, but they have occurred as rare vagrants to the south coast of England – perhaps not surprisingly, since they breed just across the Channel. This treecreeper feeds on weevils, caterpillars, earwigs and spiders and sometimes takes seeds.

This short-toed treecreeper is difficult to distinguish at a distance from the treecreeper except that the upper parts are greyer and the flanks are buff-coloured, but minor differences in mantle, flank and bill colour, and the shorter claws, can be seen at close range. The song is deeper-pitched than the treecreeper's. Like its relative it has stiffened tail feathers, which support it as it climbs up a tree trunk. It is found in both gardens and parks on the Continent.

GOLDEN ORIOLE *Oriolus oriolus*

The yellow body and black wings and tail of the male golden oriole are readily identified, but surprisingly difficult to spot amongst the foliage. However, its fluty song announces its presence. The female is green, with pale streaked underparts, and might be taken for a green woodpecker.

FOR SUCH A BRIGHTLY COLOURED bird, the golden oriole is remarkably difficult to see. It is strictly arboreal, spending most of its time high in the tree canopy, yet it is impossible to ignore the rich, fluty songs. The male is golden-yellow, with black wings and tail and a pink bill. The female is green and might be mistaken for a green woodpecker at a glance. These birds fly swiftly with long undulations, sweeping upwards to land on a branch. They rarely land on the ground and hop clumsily. Aggressive on their breeding territories, golden orioles attack trespassing crows, magpies, cuckoos and kestrels. They inhabit deciduous and mixed woodland, favouring oak, birch and willow, riverside woods of poplar, damp alder woods, orchards, parks and gardens with mature trees.

The song is a clear, pure and fluty whistle, *weela-weeoo*, which could only be mistaken for a distant mistle thrush. The song is given from high in the foliage. Migrant birds sing while on passage and in spring they are not infrequently heard in Britain. On the Continent they sing from May to July, and British breeding birds sing as soon as they arrive in mid-May. The song is heard mostly for a few hours after dawn, then intermittently through the day. Once a pair is incubating, both birds become silent; however, the fledged young are quite noisy, uttering a jay-like screeching call which their parents also use in alarm.

The nest is built suspended from the fork of a branch, with its edges woven round the branch and the main cup hanging below. Three or four eggs are laid and one brood is raised.

Golden orioles are summer visitors to Europe, arriving from April to May. They breed as far north as southern Scandinavia but rarely in Britain. They feed on caterpillars, beetles, bees, flies and spiders in the spring and summer, and in autumn cherries, dates, figs, grapes and various berries.

FACTS AND FEATURES

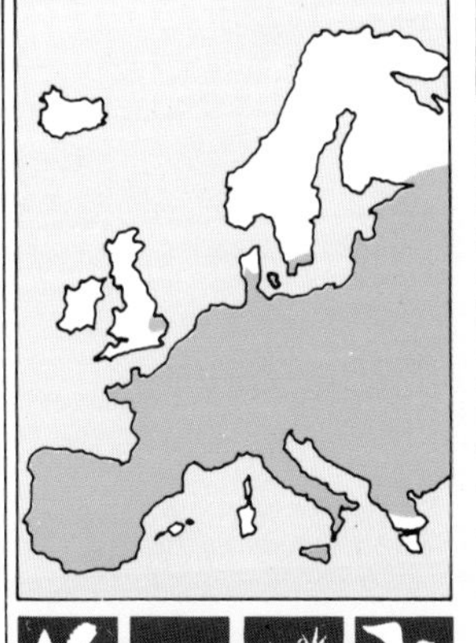

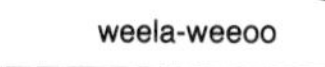

Song A clear, fluty whistle, *weela-weoo*.
Behaviour Sings from tree canopy, particularly just after dawn.
Habitat Deciduous and mixed woodland, parks and orchards, often near water.
Nest Of grass, bark and wool, suspended from the fork in a branch. Three or four eggs, one brood.
Food Beetles, caterpillars, flies, grasshoppers, spiders, and fruits and berries in autumn.

RED-BACKED SHRIKE *Lanius collurio*

THIS BRIGHTLY COLOURED species is a member of the shrike family of predatory songbirds. It has earned the nickname 'butcher bird' from its habit of storing prey impaled on thorns. The habitat includes scrub, open woodland, hedgerows, heaths, commons, orchards and gardens – anywhere with thorny bushes like hawthorn. The male is instantly recognizable with his grey head, black mask (typical of shrikes), chestnut back and white underparts with a pinkish flush. A predatory nature is made apparent by the stout, hooked bill.

Red-backed shrikes tend to perch in the open, on the tops of bushes or on tall pieces of vegetation, occasionally cocking or fanning their tails and often waving them slowly from side to side. From this vantage point a bird can see and drop onto its prey. In flight it has a rapid wingbeat and tends to undulate, sometimes gliding. It may also hover for short periods when watching for prey. On the ground this shrike occasionally hops, but rather awkwardly. Food includes beetles, butterflies, bees, grasshoppers, small mammals, amphibians, lizards, worms and also small birds and their young. Prey is taken back to a perch and held with one foot; if not eaten it may be stored, impaled on a thorn 'larder'.

The song is a quiet warbling, rather unstructured and jerky, with varied buzzing and churring notes such as *chivee-chivee, scrrr, wissi-wissi, tiu-tiu-tiu*.... It is delivered from the top of a bush or piece of scrub and is infrequent from May to July. The main call is a harsh *chack, chack* and other calls include *chee-uk*.

Nests are built in a thicket of hawthorn, gorse or bramble with five or six eggs and one brood.

Red-backed shrikes were once common in Britain, but now there are only one or two pairs remaining in England. With plenty of suitable habitat available, it is thought that dwindling food supplies have caused the decline. It may be that a change in climate, with wetter summers, has meant fewer of the large insects which this bird requires. In Europe it is found from the south up into Scandinavia; there have been records of birds breeding in Scotland, which could be a sign of colonization by Scandinavian birds. Spring arrival is in May and departure in August, with European birds migrating through the Middle East to reach their winter homes in tropical and southern Africa.

The male is a striking bird with a chestnut-red back, grey head, black face mask and white flushed pink underparts. Females have brown upperparts and may show some barring underneath. The red-backed shrike has earned the nickname 'butcher bird' as it often impales its prey on thorns or barbed wire.

FACTS AND FEATURES

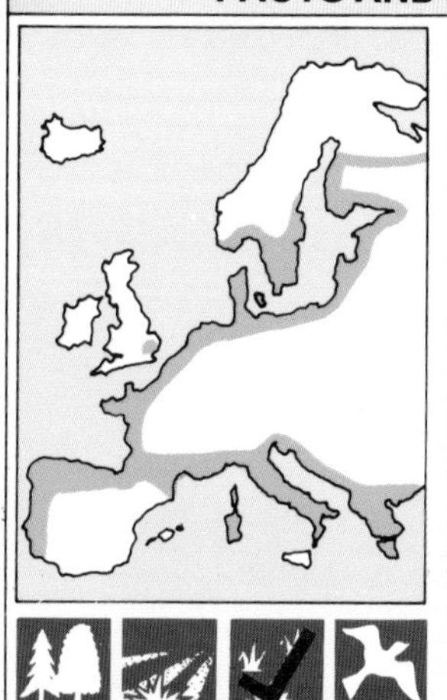

chivee-chivee, scrrr, wissi-wissi, tiu-tiu-tiu

Song A quiet warbling with buzzing notes and some mimicry.
Behaviour Sings from an exposed perch on a bush.
Habitat Hedgerows, open areas with thorny bushes.
Nest Of moss, twigs and grass in a thick bush or brambles. Five or six eggs, one brood.
Food Beetles, butterflies, bees, grasshoppers, small birds, mice, shrews, frogs and worms.

GREAT GREY SHRIKE *Lanius excubitor*

This, the largest European shrike, has the typical slightly hooked bill of all shrikes and will perch on a tree or bush to watch for prey. It eats reptiles and insects, and will pursue small birds. Its black face mask, black and white wings and long black tail contrast with the white underparts and pale grey upperparts.

THIS IS THE LARGEST European shrike and has bold grey, black and white plumage. The black face mask makes it look like a bandit waiting to pounce on a victim – and as it perches atop a tall bush, this is exactly what it is planning to do. Yet such an obvious-looking bird can be surprisingly hard to see, since if not perched prominently it may be just inside a bush, eating, or finding a suitable thorn on which to impale and store its prey. The long tail makes flight undulations fairly pronounced and the bird often glides. This shrike is a fast flier and chases after flocks of finches as it hunts. It is aggressive and reacts particularly to birds of prey, giving a loud shriek (from which the family is named) if alarmed by one. In the breeding season it frequents open woodland, both deciduous and coniferous, and also heaths, plantations, orchards and gardens. In winter it prefers open country with scattered bushes.

The song is a quiet warble of double notes interspersed with calls, squeaks and trills and even occasional imitations – *choovi-choovi, shrick-shrick, toodidi-toodidi....* The bird sings from its perch on a bush and has been heard warbling in Britain on a fine winter's day. Its call is a short, sharp *check* or *shrick* which can be repeated rapidly to form a chatter.

The nest is built in a tree or large bush and from five to seven eggs are laid in the single brood.

Great grey shrikes are widely distributed across Europe, Asia, northern Africa and northern North America. Over this large area a number of races occur with different plumages. In most of Europe they are either breeding or wintering birds. In Britain, about 100 to 200 birds spend the winter evenly distributed across England, Scotland and Wales. These wintering birds come from Scandinavia and arrive from October, leaving in March and April. They are territorial at this time and often return to the same wintering area each year. These shrikes feed on beetles, wasps, grasshoppers and butterflies as well as amphibians, lizards and small birds and mammals.

FACTS AND FEATURES

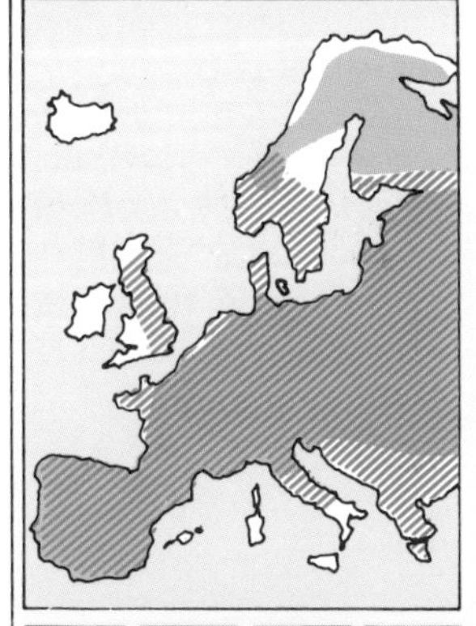

choovi-choovi, toodidi-toodidi...

Song A quiet warble, many phrases repeated twice with squeaks and trills.
Behaviour Sings from a perch on a bush or tree.
Habitat Woodland edges, heaths with trees and gardens. Open country with bushes in winter.
Nest Of grass and twigs in tree or large bush. Five to seven eggs, one brood.
Food Small birds (especially in winter) such as finches, sparrows, buntings, larks and tits. Also beetles, wasps, butterflies, grasshoppers, small mammals, frogs and lizards.

WOODCHAT SHRIKE *Lanius senator*

THIS HANDSOME red-capped bird likes open scrub with bushes, woodland edges, orchards, olive groves and gardens. It chooses higher trees and bushes to act as lookout perches than does its cousin, the red-backed shrike. Commonly perched on telegraph wires and posts by the road, it is a distinctive black-and-white bird with a chestnut-red crown and a black face mask and forehead. The back and wings are black with white scapular patches and in flight the white rump contrasts with the black tail.

Although the woodchat shrike sits out on tree and bush tops, it spends more time perched in cover than other shrikes; otherwise, it has the usual behavioural characteristics of its family. It feeds on large insects, especially beetles, butterflies, grasshoppers, bees and flies and also takes small birds such as finches, warblers and swallows. Like the red-backed shrike, it keeps a 'larder' of food impaled on thorns. A greater association with trees and its chattering calls have earned it the name 'woodchat'.

The song is a musical warbling with whistles, trills and some imitative notes, *sooi-dididi-weeeeo-chew-chew-dididi*. The song is often sustained for some time and is usually delivered from a prominent perch, but also from deep in a bush. Birds sing immediately on arriving at their breeding grounds and continue until the eggs have been laid. The call notes are varied, the most common being a jarring, chattering *schrrrrret* and a drawn-out *kschaaa* or *skeekeekeek*.

This species nests in trees, mainly on the outer branches, but will also use bushes. Five or six eggs are laid and occasionally two broods may be raised.

FACTS AND FEATURES

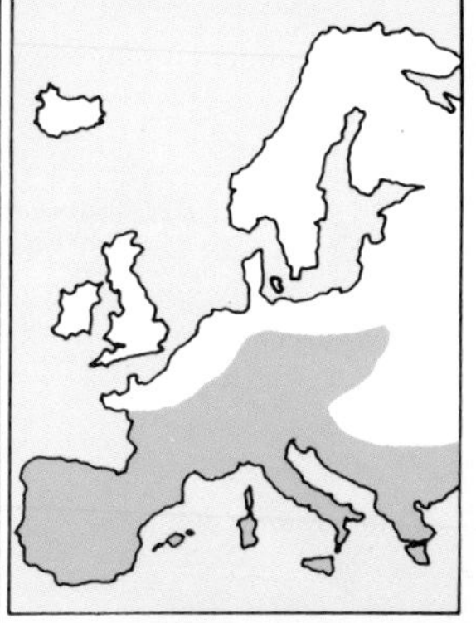

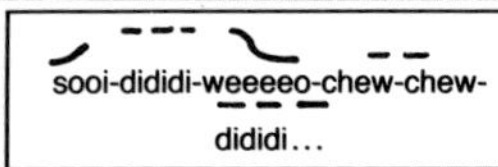

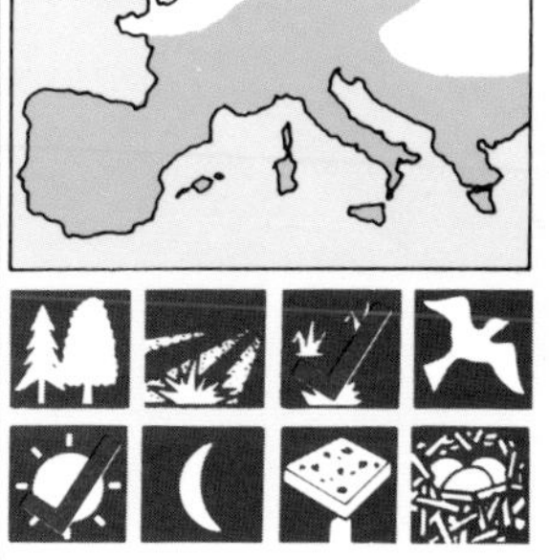

Song A rich warbling interspersed with high and harsh notes.
Behaviour Sings from a tree or bush, sometimes from cover.
Habitat Open ground with scattered trees and scrub, woodland edges, olive groves and orchards.
Nest Of roots and leaves in the outer branches of a tree or large bush. Five or six eggs, one brood normally.
Food Small birds, mainly finches and warblers, also beetles, butterflies, grasshoppers, bees and worms.

Woodchat shrikes are summer visitors to Europe and spend the winter in Africa south of the Sahara. They are found over central and southern Europe and the Near East, arriving in April and May and departing in August and September. Each spring, some birds reach Britain, but there has been no breeding.

Another mainly black-and-white shrike, found in southern Greece, is the masked shrike, *Lanius nubicus*. It has a black crown, white forehead and rufous flanks.

A distinctive bird, the woodchat shrike has a black back, wings and tail with white patches, a black face mask and white rump, and a chestnut red crown. Females have some white around the base of the bill, while young birds have a scaly brown plumage with a pale wing patch. They live in open country and can often be seen perched on fences and telegraph wires.

JAY *Garrulus glandarius*

A colourful fairly common bird in woodland, parks and gardens the jay is a member of the crow family, with a harsh screeching call. It has broad rounded black wings and shows a bold white rump in flight. Its pinkish-beige body, black-and-white crown and blue wing patches are distinctive features. It will take peanuts from bird tables and is especially fond of acorns, but, like other crows, the jay will raid the nests of small birds too.

FACTS AND FEATURES

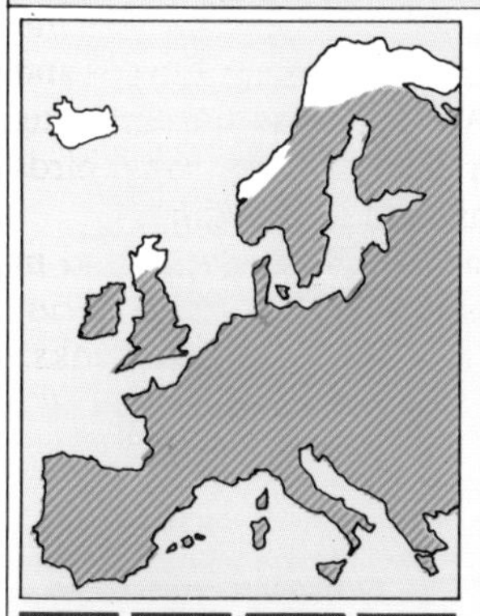

kaaa

Song Occasionally utters quiet gurgling notes mixed with harsher notes similar to the call, *kaaa, kaaa*.
Behaviour Sings quietly at spring 'assemblies' in trees.
Habitat Woodland, and hedges and gardens nearby.
Nest Of sticks, twigs and earth in trees and tall hedges. Five or six eggs, one brood.
Food Acorns, corn, nuts, beechmast, fruits and berries. Some small birds, eggs, mice, worms and insects. Takes peanuts from bird tables.

THE JAY IS one of the most colourful of European crows, found in deciduous, coniferous and mixed woodland, plantations, scrub, orchards, parks and large gardens. Oak, beech and chestnut woods are particularly popular. It is not as confiding as other crows and tends to be shy, flying off at the least disturbance. When flapping across a forest glade its black wings and tail, contrasting with its white rump, are the most distinctive feature. When perched its pinkish-brown body, black moustache, black- and white-streaked crown and blue patch on the wing make it unmistakable. In spring jays indulge in gatherings of up to 30 birds.

None of the noises a jay makes could be called a song, but it is still a communicative bird. The commonest call is a raucous screech, *kaaa* or *kraaak*, and there is also a mewing call and various harsh chuckles and croaks. Rarely the bird makes a quiet,more musical gurgling.

Nests are built in tree branches, ivy and tall hedges, or occasionally in hollow trees. Five or six eggs are laid and one brood is reared.

Jays are found across Europe and Asia in a variety of races. Continental birds, darker than the British race, are occasionally seen in Britain and Ireland. Although the species is normally sedentary, failure or exhaustion of food supplies can result in large movements.

Jays are absent from many upland areas in Britain, probably due to lack of suitable habitat such as oaks. They feed on certain insects, especially caterpillars, beetles and earwigs, and spiders. They also like nuts and fruits of various types, and have even been known to try to extract them from a hanging feeder. Large amounts of food are stored in winter and can be relocated even under snow. In this way jays play a part in the dispersal of acorns, helping oaks to spread. Large numbers of acorns and nuts can be carried at one time, thanks to the jay's enlarged oesophagus (gullet).

MAGPIE *Pica pica*

The clear-cut handsome black-and-white markings of the magpie are unmistakable. In bright light its plumage is iridescent, the black turning purple-green. It has a harsh chattering cry. Magpies will rob nests of eggs and young birds, if given the chance. They themselves usually nest well above the ground, in thickets, trees or hedges.

MUCH MALIGNED as a villain, and often regarded as a sign of bad luck, the magpie is nevertheless an attractive bird. It appears black and white, but in bright light its plumage shows a variety of iridescent sheens. The tail glimmers green and bronze, the head and neck are glossed with purple, and the wings shine blue. The long tail makes the magpie a distinctive bird whether perched or flying, and in flight its black-edged white primary wing feathers can be seen. These birds inhabit hedges, trees, bushes, thickets and woodland edges near grassland. Recently they have begun to enter suburban gardens, parks and even city centres.

This noisy and extrovert bird is difficult to ignore as it perches openly on the top of a tree or bush. The old country rhyme, 'One for sorrow, two for joy', becomes a little stretched when faced with a ceremonial gathering of magpies, from a dozen up to 50 in number. Such spectacles are thought to be part of territorial disputes involving immature birds and adults.

The harsh chattering call is the most common vocalization. A cackling *chakka-chakka-chakka* and other similarly raucous noises are delivered from the top of a hedge, but the 'song' of babbling and whistling notes is rarely heard.

Nests are usually in isolated trees, thorn hedges and thickets. They are built well above the ground and five or six eggs are laid in one brood.

Magpies are found across Europe and Asia and into North America. In Britain, they inhabit most of England, Wales and Ireland, but only parts of Scotland. These resident birds stay on their territories throughout the year. They feed on insects, especially beetle and fly grubs, caterpillars, worms, grains, fruits, nuts, small mammals, eggs and small birds. Indeed, magpies have been blamed for the loss of small birds in some areas. These crows are particularly successful at raiding nests in gardens where there is little natural protection. Faced with a ready supply of eggs and young birds, they have simply taken advantage of it. However, while magpie numbers have increased recently, they are only returning to levels they enjoyed before persecution. Magpies are often the subject of mobbing, especially in the breeding season, and they themselves mob cats and other predators.

FACTS AND FEATURES

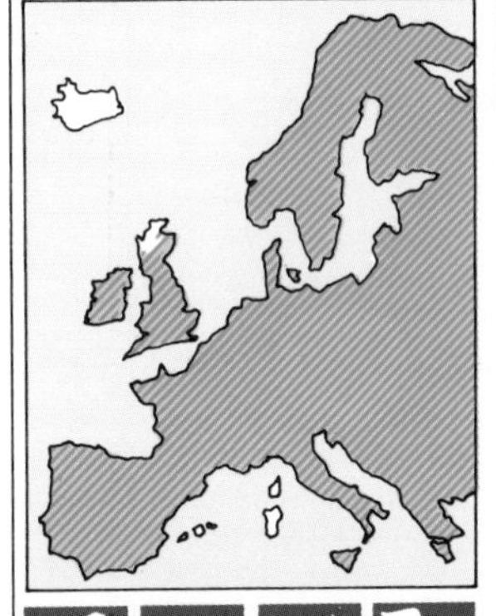

chakka-chakka-chakka

Song A quiet series of whistling and piping notes is occasionally given. Normal voice is a harsh, chattering *chakka-chakka-chakka*.
Behaviour Calls given at 'assemblies' in trees, bushes and on the ground.
Habitat Open country with hedges and trees, woodland edges and thickets.
Nest Mainly of sticks in a tall tree or bush. Five or six eggs, one brood.
Food Beetles, butterflies, caterpillars and grubs. Also small mammals, birds, eggs, worms, wheat, fruits and nuts. Visits bird tables.

CHOUGH *Pyrrhocorax pyrrhocorax*

FACTS AND FEATURES

kee-ah

Song Usual call note is a clear *kee-ah* and some birds give a chattering twitter.
Behaviour Calls in flight and when perched.
Habitat Coastal cliffs and pastures in Britain, mountainous areas elsewhere in Europe.
Nest Of sticks, stalks and grass in a rock crevice or on building. Three or four eggs, one brood.
Food Fly and beetle grubs, some caterpillars and worms.

THE CHOUGH looks a typical all-black crow, but it is far from a typical member of the family, being distinguished by its long, down-curved red beak and red legs. In Britain and Ireland it is found along western coasts with cliffs and pasture, and a few birds live inland in Wales. It avoids 'improved' pasture, preferring traditionally grazed turf, which now confines its last strongholds. This master aerobat flies effortlessly in the upcurrents of air along cliff faces, swooping and diving or simply gliding with the tips of its wing feathers separated in a distinctive silhouette. On the ground the chough walks and runs in search of food and also hops when hurried. Elsewhere in Europe it is found in mountainous areas inland, where its calls echo between the faces of rocky valleys.

The call is a ringing *kee-ah* or *chee-ow*, given from a perch on the ground or cliff ledge and also in flight. On the Continent, birds have been heard to give a chattering twitter.

Nests are built in the rock crevices and caves of coastal and inland cliffs, quarries and sometimes ruined buildings. Three or four eggs are laid and one brood is reared.

Choughs are found across Europe, Asia and North Africa. They are residents and do not move far from their breeding areas. Occasionally birds have been seen several hundred kilometres from the nearest breeding sites, but this is exceptional. They feed on fly and beetle grubs, worms and ants, which they dig up with their long bills. When the ground is frozen hard they may turn to barley and oats. Because of their mainly insectivorous diet, choughs are vulnerable to cold winters. They have been known to forsake the freezing coastal pastures and enter suburban gardens.

A similar species, which overlaps with much of the chough's distribution in the mountains of Europe, is the alpine chough, *Pyrrhocorax graculus*. It differs in having a shorter, yellow bill and its voice is quieter – usually a whistling *chirrip* and some crow-like notes.

The down-curved red bill is an unusual feature of this member of the crow family, but its all-black plumage is more typical. It has well-separated primary wing feathers which project like fingers when it is in flight and help to distinguish it from the similar-sized jackdaw. The chough is a good flier and will glide and sail effortlessly along coastal cliffs and mountain rock faces, uttering its loud, ringing call.

JACKDAW *Corvus monedula*

THIS SMALLEST OF THE black crows has a distinctive grey nape and pale grey eye. Its size readily distinguishes it from its frequent companions, rooks and carrion crows. The jackdaw nests in woodland, parks, hedges and gardens, on sea cliffs, inland rock faces and quarries, and in places has successfully adapted to an urban environment. It is commonly found among flocks of crows, feeding on pasture or freshly ploughed fields and flying to large communal roosts, in woods and plantations, at the end of the day. On the ground this bird walks quickly with a jaunty air, and in flight it has a faster wingbeat than the larger carrion crow and can twist, turn and dive.

The two main calls are *keeaw* and *tchack*, both having been suggested as origins for its name. Flocks of jackdaws and birds in breeding colonies indulge in outbursts of calls in loud chorus. A low rippling 'song' incorporating some calls has also been heard.

Jackdaws nest in holes and crevices in rocks, trees and buildings, and even in rabbit burrows, chimneys in urban areas (sometimes blocked by a nest) and large hole-fronted nestboxes. Larger nest sites are filled with twigs – one nest in a loft was a pile of sticks several feet high, nearly reaching the entrance hole in the roof. These birds often breed communally in neighbouring trees and along cliff faces. Between four and six eggs are laid in one brood.

Jackdaws are found throughout Britain and Ireland, where they are resident. Young birds may move from their home areas in autumn, and there is a winter migration with birds from upland areas moving south and west. In severe winters large flocks have been seen travelling through Cornwall and south-west Ireland. Continental birds from Scandinavia arrive on the east and west coasts of Britain in October and November, and return from March to May.

The diet consists of beetles, caterpillars, grubs, flies, snails, small mammals, frogs, young birds and eggs, with grains in winter. Along with other crows, jackdaws scavenge around farms, towns and rubbish tips, readily entering gardens to take advantage of bird table scraps. They are often represented in literature as 'thieves', since they pick up small shiny objects and take them to their nests.

FACTS AND FEATURES

Song Sometimes gives a song of rippling notes. Call notes are *keeaw* and *tchack*.
Behaviour Vocal in flight and when perched.
Habitat Farmland, parks and cliffs.
Nest Of sticks in a tree cavity or other hole. May nest in buildings, chimneys and large open-fronted nestboxes. Four to six eggs, one brood.
Food Beetles, caterpillars, flies, small birds, eggs, mice and frogs, some cereals and fruits. Feeds at bird tables.

The smallest black crow, the jackdaw has a grey nape and pale grey eyes. It will nest in gardens, using large nestboxes, but is equally at home in an old chimney. Jackdaws can be bold and will visit gardens for food, running and hopping along the ground. They roost communally with other crows, forming large mixed flocks.

ROOK *Corvus frugilegus*

The bulky nests of rooks – large black crows with pale bills and shaggy leg feathers – are a common winter sight at the tops of groups of tall trees, where they nest in colonies. They feed mainly in fields on soil pests, worms and grain. A rook taking food to its young will often show a bulging pouch of food beneath its beak.

THE 'BARE-FACED CROW' is a descriptive local name for this familiar bird. Large and black, it has a bare area of pale grey skin on its face and a pale bill. Its plumage is similar to a carrion crow's, but its leg feathers are long and shaggy. Rooks are highly gregarious and breed in large colonial rookeries, unlike the carrion crow, which is a solitary nester. This has led to the belief that a single 'rook' must be a crow and a flock of 'crows' are rooks, which is not always so. Rooks form flocks with both jackdaws and carrion crows and roost communally with them. Roosts may be at the site of a rookery and sometimes in winter birds join a central roost from many surrounding rookeries. In late February these large roosts have broken up and birds return to their own rookeries to commence breeding.

The most familiar call is a loud *caw* or *kaah*, pitched higher than the call of a carrion crow. Roosting or nesting birds are always conspicuously noisy. They also give a gull-like *ki-ook* and have been heard to make chirruping and spluttering noises at the nest.

The favourite sites for nests are the tops of tall elm trees. A rookery may consist of many hundreds of pairs of birds, and one British colony numbered nearly 7,000. Where there are no suitable tall trees, nests can be sited in bushes. From three to five eggs are laid and one brood is raised.

Rooks are found all over Britain except for the uplands of Scotland and Wales. The population has experienced decreases in recent years, attributed to the use of pesticides and changes in agricultural practice. In Europe, these birds are absent from northern Scandinavia and are winter visitors to southern Europe. While British birds do not move far from their breeding areas, many continental birds migrate south and west in winter. Rooks are omnivorous, and fly in a straight line from feeding area to roost, hence the phrase 'as the crow flies'.

FACTS AND FEATURES

kaah, kaa-aah

Song Sometimes gives chirruping and churring notes but usually a hoarse *kaah*.
Behaviour Calls when perched and in flight. Most vocal at rookeries.
Habitat Agricultural land with trees.
Nest Of sticks in tall trees. Nests colonially. Three to five eggs, one brood.
Food Cereals, roots, fruits, nuts, beetles, grubs, caterpillars and worms.

CARRION CROW *Corvus corone*

Two distinct races of carrion crow exist. One is all black, while the other, the hooded crow, has a grey back and belly with a black head (hence the 'hood') and black wings. Both races interbreed to form hybrids with intermediate plumage. Carrion crows can be distinguished from rooks by their black bills and lack of feathering around their legs. Unlike rooks they are solitary breeders.

UNLIKE ITS RELATIVES the jackdaw and rook, the carrion crow tends to be more of a loner in the breeding season. It differs from the rook in having a completely feathered face and a black bill, and its all-black plumage has made it an object of fear and loathing among many. A liking for corpses and eggs led to continuing persecution, but despite such pressure the carrion crow remains a common and widespread bird. It is found on farmland, moors, heaths and coastal areas, in open woodland, parks and gardens.

The call is a hoarse, croaking *kraaa*, given from a perch and in flight. Other calls include a high *keerk*, and a *honk* rather like an old motor horn. Croaking and crackling types of 'song' have also been heard.

The nest is built high in the fork of a tree, in a bush or heather, or on a cliff ledge. Buildings and even electricity pylons are also used. From four to six eggs are laid and one brood is reared.

The carrion crow is found across Europe and Asia. The main race occurs in France, Germany, Spain and England as well as eastern Asia. In Ireland, northern and western Scotland, Scandinavia, eastern Europe, Italy, Greece and western Asia a distinct race of this species is found. It is called the hooded crow and differs markedly from the other race of carrion crow in having a grey body with a black head and wings. Where the two races overlap hybridization occurs, producing intermediate forms. While the carrion crow race is resident and non-migratory, hooded crows from northern Scandinavia migrate south and west, reaching Germany, the Netherlands, France and eastern Britain in the winter.

Carrion crows feed on carcases and their association with sheep carrion has led to accusations of attacking lambs. They also take young birds and eggs, which has resulted in attempts by gamekeepers to control their numbers. Small mammals, frogs, beetles, worms, molluscs, grains, nuts and seeds are all eaten.

FACTS AND FEATURES

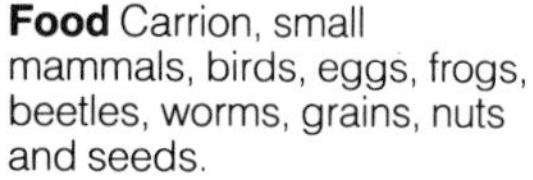

Song An occasional croaking song with notes resembling the hoarse *kraa* call.
Behaviour Calls when perched and in flight.
Habitat Farmland, woodland, moors and coasts with trees.
Nest Of twigs, sticks, earth and moss in a tree fork, bush or cliff ledge. Four to six eggs, one brood.
Food Carrion, small mammals, birds, eggs, frogs, beetles, worms, grains, nuts and seeds.

RAVEN *Corvus corax*

The largest of the crows, the raven has a wide wingspan, long wedge-shaped tail and a massive deep black bill. In Britain it is mainly confined to mountainous and coastal areas. It eats carrion, but will also kill rabbits and small birds.

THE LARGEST OF THE CROW FAMILY, and the largest European songbird, this large, black bird has predictably been looked on as a source of evil – yet it has also been held to bring good luck on occasions. Many myths surround the raven. For example, it was thought to leave its young at the nest to starve, which gave rise to its collective name, an 'unkindness' of ravens. In flight, ravens differ from rooks and carrion crows by having a wedge-shaped rather than a square end to the tail. Their wingbeats are slow and they both glide and soar. They are adept at aerial acrobatics and a pair often dive and wheel together during the breeding season.

Ravens are found in coastal and upland areas, along sea cliffs, on moorland and by rocky inland cliffs. They are thought of as solitary, but where they are numerous they may feed and roost in flocks.

The most familiar call is a deep *prrrk* when perched or in flight, and also rattling and bubbling noises.

Nests are built mainly on cliff ledges. Occasionally they are sited in trees, which was common practice when they were plentiful in Britain, and is still seen today on the Continent. Nesting sites are often traditional and may be alongside peregrine falcons. Breeding commences in February and between four and six eggs are laid in a single brood.

Ravens are found over a large part of Europe, Asia, North Africa and North America. They are the only members of the crow family to breed in Iceland. In Britain, the species is found only in the west and north – it was once plentiful throughout lowland England, but disappeared through persecution. It was also exterminated in areas of France, Germany, Belgium and the Netherlands. Ravens are resident in Europe and stay near their breeding territories. They are omnivorous, eating mainly sheep carrion and small mammals, birds, eggs, beetles and some nuts and seeds. They also scavenge scraps from rubbish tips.

FACTS AND FEATURES

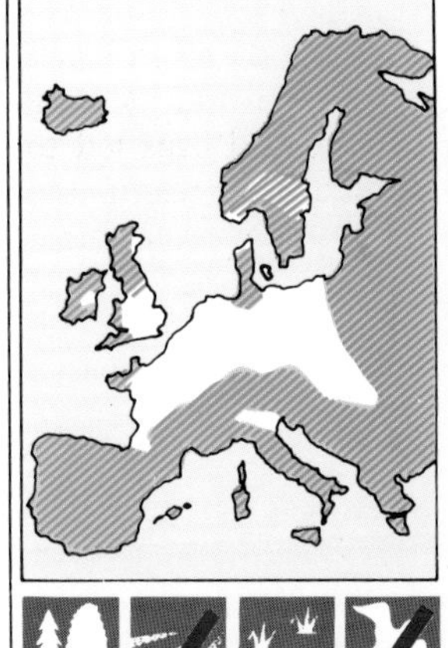

prrrk, prrrk

Song A variety of rattling and bubbling notes but the usual call is a deep *prrrk*.
Behaviour Calls when perched and in flight.
Habitat Mountainous and hilly areas and coasts. Woodland on the Continent.
Nest Of sticks, moss and leaves on a cliff ledge or in a tree. Four to six eggs, one brood.
Food Carrion, small mammals, birds, eggs, beetles, nuts and seeds.

STARLING *Sturnus vulgaris*

FACTS AND FEATURES

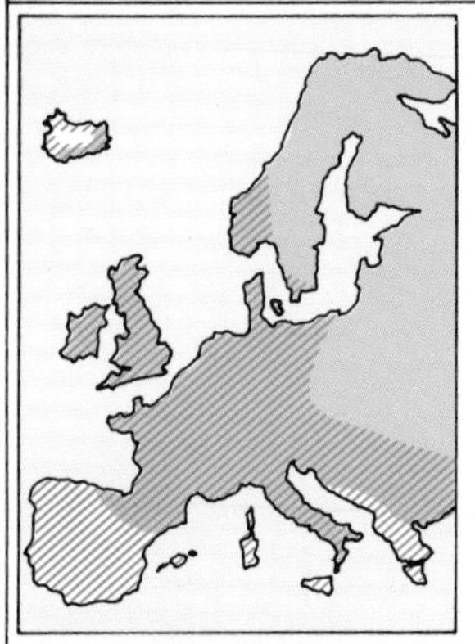

schwer-schwer-schwer, durchicki-durchicki, tip-tip-tip, seeooo

Song A variety of warbling, chirruping and whistling notes with much mimicry. A common phrase is a high, descending *seeooo*.
Behaviour Sings from an elevated perch on a tree or building, often flicking wings out.
Habitat Many urban and rural areas, except the most desolate.
Nest Untidy mass of straw in a hole in a tree or building. Uses large hole-fronted nestbox. Five to seven eggs, one or two broods.
Food Beetles, grubs, spiders, worms, caterpillars, nuts, seeds and berries. Often dominates bird tables, eating almost anything.

A LARGE FLOCK OF STARLINGS preparing to roost is an impressive sight, the birds packed tightly together in a mass yet all wheeling around simultaneously. Starlings are at home in both the open countryside and the city centre. Many large roosts have been established in central London, where birds huddle together on window ledges, safe from predators and warmer than they would be outside the city. They are often looked on as bullies, taking food from other birds and always squabbling. Yet they are attractive birds and in breeding plumage, under bright light, their black feathers have a purple and green sheen. The yellow bill has a bluish base in the male, reddish in the female.

The song consists of an endless variety of warbling, chirruping and whistling notes. One common phrase is a high, descending *seeooo* followed by a series of *wee-wee-wee-wee-wee* notes. The starling is a superb mimic and can copy anything from a curlew to a blackbird, even imitating mechanical noises like ringing bells. When singing it perches high on a roof or tree and fluffs out its throat feathers, holding the head up and flicking out the wings as it repeats phrases. These birds sing all year, but may become quieter in July. A grating *tcheer* is the usual call, but elements of the song can be used as single call notes.

The natural nesting site is in a tree hole or rock crevice. However, starlings have adapted to using holes in buildings and readily take to hole-fronted nestboxes.

Found across Europe, the starling is a summer visitor to most of Scandinavia, where it is regarded as a herald of spring – much as the swallow is in Britain. These Scandinavian birds migrate south and west in autumn and many come to Britain, adding to our own already large population. Starlings feed on beetles, weevils, flies, flying ants, earwigs, grubs, caterpillars, worms, spiders, nuts, seeds and berries. They eat virtually anything from a bird table, often fighting off all other species.

In Spain and Portugal, the starling is replaced by the similar spotless starling, *Sturnus unicolor*. This species is unspotted in the breeding season. Its voice is similar to a starling's but has more whistles.

A brash and cheeky bird, the starling has iridescent purple and green plumage and a yellow bill. Flocks of starlings can descend upon bird tables and remove the food before other birds get a chance. It has taken to roosting in town centres, where the temperature is a little warmer than the surrounding countryside. Large flocks will gather together noisily before flying off to the roost.

TREE SPARROW *Passer montanus*

A quieter relative of the house sparrow, the tree sparrow is also smaller, with an all-chestnut crown. Its black bib is also smaller and its white cheeks have a characteristic black spot. Males and females are identical. Although it will nest in artificial boxes it rarely comes close to buildings.

FACTS AND FEATURES

— — —
Chip, chip, chip

Song A series of chirping notes, higher than a house sparrow's song.
Behaviour Sings and calls from trees or on the ground.
Habitat Woodland edges, trees and gardens in Britain. Also buildings in parts of Europe.
Nest Of straw in a tree hole, thatch, roof or hole-fronted nestbox. Four to six eggs, two or three broods.
Food Seeds, corn, beetles, caterpillars, aphids and spiders.

A SMALL AND DAINTY BIRD, this species differs from the house sparrow in its chestnut crown, black cheek spot and pale collar. It is also slimmer, quieter and more secretive and is often overlooked. In Britain, the tree sparrow is found near wooded areas, old orchards, parkland and hedgerows; it does not associate much with habitation, although it may be found around farm buildings. In parts of Europe and farther east it has a similar lifestyle to the house sparrow.

The song is slightly more structured than a house sparrow's, but still has chirping notes such as *chip, chip, chip, chittup-chirrtoowhit, chittup-chirrtoowhit.* This bird is not very vocal and is heard infrequently from March to May. The call is a *chip* higher than a house sparrow's, and in flight it gives a *teck, teck*.

The untidy nest is sited in a hole in a tree, cliff, quarry or building and occasionally in a sand martin's burrow. Sometimes nests are built in the branches of trees and shrubs or in the large nests of other birds. The tree sparrow may use a hole-fronted nestbox if the hole is too small for a house sparrow (about one and a quarter inches or three centimetres in diameter). Where there are sufficient breeding sites it may form colonies. It lays between four and six eggs and has two or three broods.

The tree sparrow is absent from northern Scandinavia and parts of southern Europe. In Britain, it is scarce in northern and western Scotland, western Wales and Cornwall, and is scattered thinly around the Irish coast. This resident moves little in winter, although birds that breed close to the Arctic Circle migrate south. Individuals seen on the east coast of Britain in autumn could be birds from the Continent. Tree sparrows form winter flocks which may contain several hundred birds. They also join mixed flocks of chaffinches, greenfinches, buntings and yellowhammers, feeding at the edges of ploughed fields and in short, arable vegetation. Food consists of grains, grass and weed seeds, beetles, caterpillars, aphids and spiders. Birds may come to bird tables for a seed mixture in winter, but this is uncommon.

HOUSE SPARROW *Passer domesticus*

This small noisy bird has adapted successfully to living alongside man. On close inspection the males are attractive, with grey crowns, chestnut head, white cheeks and black bib, black-and-brown patterned wings. The females are duller with indistinct wing bars. Sociable birds, living in large flocks, sparrows will breed in almost any suitably sized hole and will also use nestboxes, often taken over from blue tits.

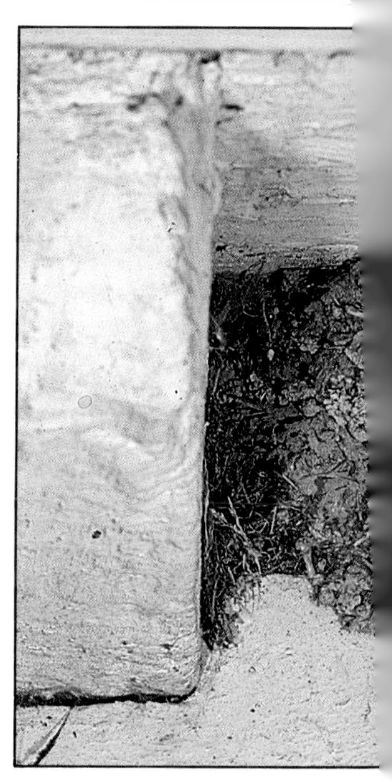

FACTS AND FEATURES

Chirrup, chrrp, chup

Song A chirping song of eight to 10 notes.
Behaviour Song often given when many males are together, on the ground or in a bush or tree.
Habitat Urban areas and rural farmland near buildings.
Nest Of straw, untidily domed with a side entrance, in a hole in a building, ivy, a bush, a martin's nest or a hole-fronted nestbox. Three to five eggs, three broods common.
Food Corn, seeds, fruits and buds, also insects and worms. Takes peanuts from feeders and any other scraps.

THIS UBIQUITOUS BIRD is not quite as domestic as its name implies, but it has adapted successfully to live alongside people. It is a cheeky, chirpy bird found in every urban and rural environment where humans are present. It has taken advantage both of nesting sites in buildings and of the ready food sources that agriculture and waste disposal provide. When studied closely, the male is rather handsome with a grey crown, chestnut head, white cheeks and a black bib. His black-and-brown patterned back and wings are broken by a white wing bar, and there is a grey rump and brown tail. This active, perky sparrow hops on the ground, often flicking its tail as it searches for food, but it also perches freely on buildings and in trees. House sparrows are sociable birds, often gathering in large flocks to feed and roost. They always live close to people, in cities, towns, suburban gardens, parks, farmland and even isolated upland areas.

The song is a noisy chirping of some eight to 10 notes like *chip*, *chirrup*, *chrrp* and *chup*. When a flock of birds assembles their excited din can be quite deafening. They sit and sing on roofs, trees or the ground, or groups of chattering males chase a single female, vying for her attention vocally. Males are noisiest from February to July. Calls are varied and include a loud *chee-ip* and a rattling twitter.

The untidy nest is built in almost any kind of hole – in a building, thatch, ivy, tree or hedge, in the large nest of another bird such as a rook or heron, and even in a house martin's nest. A hole-fronted nestbox is often used, to the detriment of tits. From three to five eggs are laid, and two or three broods may be reared.

The species is found naturally across Europe and

Asia, and in North Africa. It has also been introduced to almost every part of the world, with an estimated global population of about 500 million birds. It is a resident, staying close to its breeding site, although it may move from exposed upland areas in winter. The thick bill indicates a seed-eater, but the house sparrow takes virtually anything, from corn and seeds to insects, fruits and worms, and flocks visit bird tables for scraps of all kinds. They copy tits in taking cream from milk bottles and have even mastered the tits' technique of clinging to peanut feeders.

In the Mediterranean lives a close relative, the Spanish sparrow, *Passer hispaniolensis*. It has an all-chestnut crown and a more extensive black bib than the house sparrow, spreading down the flanks. Its calls are similar. Hybrids between the two species, called 'Italian' sparrows, resemble the house sparrow but with a chestnut crown.

They often take over house martins' nests (left) *before they have been completed.*

Partial to nuts, the house sparrow will habitually come to feeders (above). *To prevent over use by this species, the holder should be suspended from a piece of wire.*

A typical example of a nest (right) *placed in the roof of a building.*

ROCK SPARROW *Petronia petronia*

This southern European species resembles a female house sparrow but has a more strongly striped head and a yellow breast spot. It has a broad pale stripe over the eye and a buff central crown bordered with darker brown. Living in mountainous areas, it is more generally shy and wary than the house sparrow.

A SQUAT-LOOKING BIRD, similar to a female house sparrow, the rock sparrow has a broad, pale supercilium ('eyebrow') and a buff central crown stripe bordered by darker brown. Its breast is well streaked, and in flight it may show the white spots at the end of the tail. When the male is singing or displaying, his yellow breast spot becomes visible. Rock sparrows inhabit hilly and mountainous regions, preferring rocky slopes and cliffs but also living in farmland by buildings, bridges and ruins. They spend much time on the ground and tend to run like a pipit rather than hopping like other sparrows.

The song is a hissing or wheezing *vi-viep* or *weel-eep*. It is uttered persistently and often forms a noisy chorus with other birds which echoes across a rocky valley. The call note is a high, nasal *pey-i*. In flight it has chipping notes, similar to but more musical than those of a house sparrow.

The nest is built in a hole in rocks or in a wall, hollow tree or bank; the nest of a house martin may be used. Rock sparrows sometimes form mixed colonies with house sparrows. Five or six eggs are laid, with two broods often raised.

Rock sparrows are found in southern Europe, where they are resident and stay near to their breeding areas. They may move to lower ground in winter, when they mix with flocks of house sparrow and finches, feeding on farmland. They eat caterpillars, grubs, small grasshoppers and other insects in summer and turn to grass and weed seeds, berries and fruits in winter. Despite their sedentary nature there has been one record of a rock sparrow in Britain, several hundred kilometres from the nearest breeding birds.

A European species related to the sparrows is the snow finch, *Montifringilla nivalis*. It has a grey head, brown back, black throat and whitish underparts. In flight it reveals startling black-and-white wings and tail. This bird breeds in rocky areas on high mountains. Its song is a monotonous *ti-witchu, ti-witchu* and its calls are a short *pshee* and a *prrrrt* of alarm.

FACTS AND FEATURES

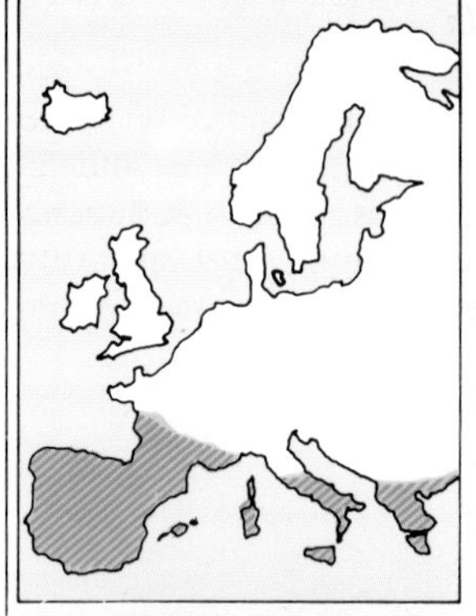

vi-viep

Song A hissing or wheezing *vi-viep*.
Behaviour Male displays his yellow throat spot when singing from a rock.
Habitat Rocky, mountainous areas, ruins and other buildings.
Nest Of straw, grass and roots in a hole in rocks, a wall or tree. Five or six eggs, two broods.
Food Caterpillars, grasshoppers and other insects in summer; seeds and berries in winter.

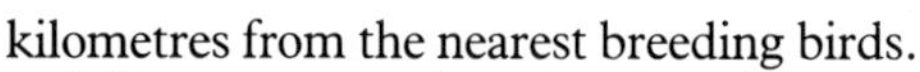

CHAFFINCH *Fringilla coelebs*

THIS COLOURFUL FINCH is a contender for the title of commonest British bird, vying with the blackbird and wren. The male chaffinch is unmistakable with his pinky-red breast, grey head, and chestnut back with a green rump. Most striking in flight are the white shoulder patch and wing bar and the white outer tail feathers. This bird lives in mature deciduous woodland, farmland, parks and gardens. It is tame and confiding, readily visiting city parks and suburban gardens and becoming virtually resident at a bird table. It walks or hops along the ground with a jerky motion, and like all finches its flight is undulating. The chaffinch is territorial in summer but flocks with others of its kind in winter, when it may join bramblings and greenfinches, feeding in fields and roosting in evergreens. It can be aggressive and often attacks its reflection in a car's hubcap or wing mirror, even battering against a window pane in an attempt to vanquish its unreal rival.

The chaffinch's song is one of the joys of spring, ringing out in cheerful exuberance. It is a loud, accelerating series of notes finishing with a flourish at the end: *chip-chip-chip-chwee-chwee-tissi-chooeeo*. It lasts for two or three seconds and at the start of the breeding season a male may sing more than 3,000 times a day. Chaffinches use song posts on trees and tall bushes within their territories. They commence singing in mid-February, continuing until early July, with occasional song at other times. The phrasing of the song and quality of the notes can vary between individuals. Indeed, in isolated areas where populations are separated by hills or mountains, it may be possible to identify chaffinches from particular places by their 'dialects'. They also have at least 14 different calls, especially a loud *pink-pink* or *chink-chink*, a clear *wheet* given in spring, and a low *tupe* in flight.

Nests are built in hedges, bushes and small trees, hidden in forks. They are well camouflaged and only one brood of four or five eggs is laid.

Chaffinches are widespread and resident throughout Britain and Ireland. Birds from Scandinavia and farther east migrate to Britain in the winter, and it has been estimated that some 10 to 20 million of them come here. The diet varies during the year. In spring and summer they eat flies, moths, beetles, caterpillars and spiders. In autumn and winter they feed on beechmast (beech nuts), cereals and weed seeds. In the garden they eat seed mixtures from ground or bird table.

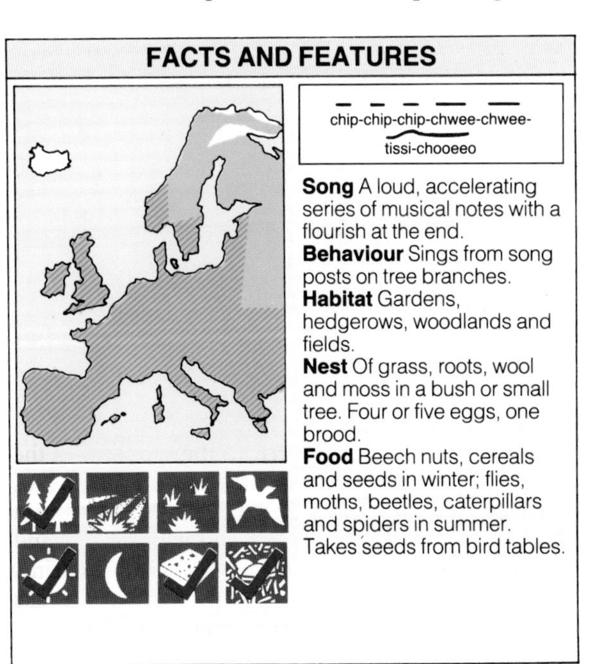

FACTS AND FEATURES

Song A loud, accelerating series of musical notes with a flourish at the end.
Behaviour Sings from song posts on tree branches.
Habitat Gardens, hedgerows, woodlands and fields.
Nest Of grass, roots, wool and moss in a bush or small tree. Four or five eggs, one brood.
Food Beech nuts, cereals and seeds in winter; flies, moths, beetles, caterpillars and spiders in summer. Takes seeds from bird tables.

Perhaps the commonest European finch, the chaffinch is unmistakable with the pinky-red breast, chestnut back with green rump and grey head of the male (the female is browner). In flight its most distinctive features are the white wing bars, which show clearly, and the white outer tail feathers. The loud song, delivered from regular song posts, is one of the heralds of spring.

BRAMBLING *Fringilla montifringilla*

THIS NORTHERN COUSIN of the chaffinch is similar to its relative in shape but has less white on the wing and a white belly. Its breast is orange-buff and in spring the male's head and upper back are black, becoming mottled with buff in winter. In flight it displays a narrow white rump and does not have the chaffinch's white outer tail feathers. Also like its relative, the brambling is territorial in the breeding season, but forms flocks in winter. It breeds in northern birch woods, open conifer forest and tall willow scrub, and spends the winter feeding in open fields and woods and roosting in conifers. This bird is also known as the bramble finch, although it has little to do with brambles except occasionally roosting in them. In spring and summer it spends most of its time in trees, feeding and nesting, while in autumn and winter it forages mainly on the ground.

The brambling's song bears no relation to that of the chaffinch. It is a disappointing *dsweea*, rather like the song of a greenfinch. The distinctive call is a harsh *tsweek*, which often identifies the presence of this bird in a large, mixed flock of finches. In flight it has a hard *chik* call, rather like a chaffinch.

The nest is built quite high in a tree. Six or seven eggs are laid; in the south of its range the brambling may have time to raise two broods, while farther north it is restricted to one.

Bramblings completely vacate their Scandinavian breeding areas in winter, migrating south and west to cover the rest of Europe. The extent of their migration and the numbers arriving in different parts of Europe depend on food supplies, and the availability of beech nuts (beechmast) plays a major role. The numbers of birds visiting Britain in winter may vary from 50,000 to two million. In 1981, when large numbers arrived, an estimated 150,000 birds used one roost in Merseyside. On the Continent, roosts containing millions of birds have been recorded. Some bramblings remain here into the summer and there has been occasional breeding in Scotland. In spring they consume insects. In winter they eat beechmast and weed seeds, and also visit bird tables for seeds, sometimes attempting to reach peanuts in hanging feeders.

FACTS AND FEATURES

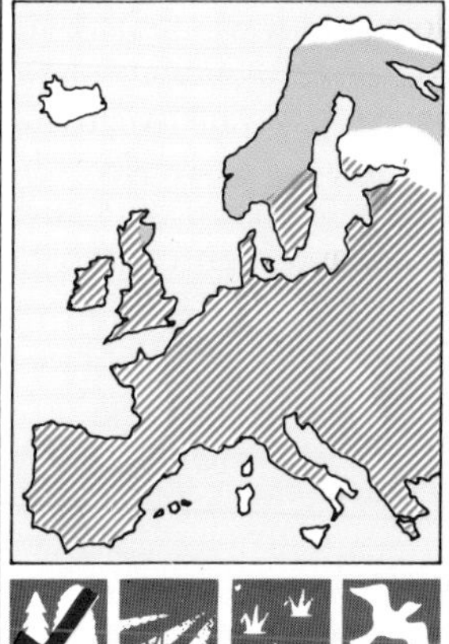

— —
dsweea, dsweea...

Song A prolonged, greenfinch-like *dsweea*.
Behaviour Sings from song posts in trees.
Habitat Breeds in birch and mixed woodland, winters in fields and woods.
Nest Of grasses, bark and lichens in a tree. Six or seven eggs, one or two broods.
Food Seeds, beech nuts, caterpillars and insects. Visits gardens for seeds in cold winters.

A northern finch, the brambling is seen elsewhere in Europe during the winter. In winter the male has a greyish head, which becomes black in the spring. In flight it shows a white rump, which contrasts with its black tail and wings. The female is similar in appearance. It feeds with other finches in winter flocks, sometimes visiting garden bird tables for seeds.

SERIN *Serinus serinus*

THIS TINY BIRD is the smallest finch in Europe. The male is distinctive with his bright yellow head and breast, white belly and yellow rump, which in flight contrasts with the streaked back. It is difficult to overlook a serin in spring since it is always active, flying from one perch to another. These birds have the characteristic bouncy flight of a finch and this, combined with their small size and high call, make recognition reasonably easy. Serins are found along woodland fringes, vineyards, orchards, hedgerows, parks and gardens. The species is a close relative of the wild canary, from which the familiar cage birds have been bred, and many of its sounds are similar to those of a canary.

The song is a monotonous, jingling *seeteeseeteeseetee*, delivered from a perch or in a greenfinch-like song flight with a slow flapping wingbeat and circling action. It is heard throughout the year, although less from August to September. The call note is a drawn-out *tsooee*, similar to that of a greenfinch, and in flight the bird gives a rapid, twittering *si-twi-twi-twi*.

The nest may be built on the branch of an oak or pine tree or in thick bushes, bramble or ivy. Four eggs are laid and two broods can be reared.

Serins inhabit most of southern and central Europe and have been spreading north from their Mediterranean stronghold for some time. In Britain they have bred occasionally over the last 20 years, with about six pairs nesting in a good year. Apart from such isolated breeding records the species is a scarce visitor to Britain, usually turning up along the south coast in spring. However, recent records indicate that its colonization of Britain may become more permanent and widespread, given time. It is a resident in southern Europe but a summer visitor farther north, migrating to the Mediterranean in winter. This finch feeds on buds, catkins, weed and grass seeds and some insects.

A European relative of the serin is the citril finch, *Serinus citrinella*, which is confined to mountainous regions. It looks like a miniature greenfinch with a grey neck and has a song reminiscent of both the serin and goldfinch. The call note is a plaintive *tsi-ew* and the flight call a bouncy *chitt-itt-itt*.

A tiny yellow finch with a high, jingling, rapid song. It is commonest in southern Europe, where it is found in gardens and parks. It has a brown-streaked back, streaked flanks and a bright yellow rump, which shows in its characteristic bouncy flight. Its song and appearance can be likened to that of a small canary, to which it is related. The female's plumage is not as yellow as the male's.

FACTS AND FEATURES

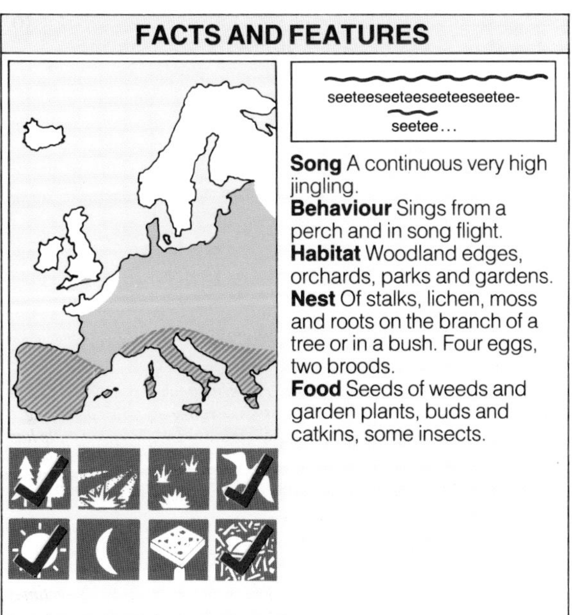

Song A continuous very high jingling.
Behaviour Sings from a perch and in song flight.
Habitat Woodland edges, orchards, parks and gardens.
Nest Of stalks, lichen, moss and roots on the branch of a tree or in a bush. Four eggs, two broods.
Food Seeds of weeds and garden plants, buds and catkins, some insects.

GREENFINCH *Carduelis chloris*

FACTS AND FEATURES

tsweeee

Song A drawn-out, nasal *tsweeee* is the main note, combined with other twitterings.
Behaviour Sings from a perch at the top of a tree or in song flight.
Habitat Open woodland, gardens, thick hedges, plantations and fields.
Nest Of twigs and moss in a hedge, evergreen or bush. Four to six eggs, two broods.
Food Cereals, weed seeds, yew berries, blackberries, hips and some insects. Readily feeds on peanuts and sunflower seeds at bird tables.

THIS CHUNKY BIRD is one of the most familiar garden finches, especially at a bird table. The male is stout-billed, olive-green above and yellow-green below, with a yellow stripe along the edge of the closed wing. The tail is forked with yellow sides. The greenfinch is a gregarious bird, often forming winter flocks which feed and roost together. Just before dusk in winter, bands of greenfinches, chaffinches, goldfinches, siskins and bramblings all collect in tree tops near a roosting area, which may be a small plantation of thick bushes such as rhododendron. Greenfinches are generally found in open woodland, plantations, hedgerows, parks and gardens. The flight is undulating and the bird hops when feeding on the ground. It can be aggressive over food, opening its beak in a threatening manner if a bird comes too close.

The song is not very musical, consisting of a drawn-out, nasal *tsweeee* mixed with twittering phrases. It can be given from a perch at or near the top of a tree, or from a bush, or in a characteristic song flight. In this flight the male circles with slow, exaggerated wingbeats, often changing direction erratically. The main song period is from early March to late July. In flight there is a twittering *chichichichichit* call and other calls are *chup*, *teu* and *tsooeet*.

The nests are well hidden in thick cover, such as evergreens and bushes like hawthorn and elder. There are from four to six eggs and two broods, with some clutches being laid as late as mid-August.

Greenfinches are resident in most of Europe but are summer visitors in parts of Scandinavia. British birds may move some distance in the winter, but they rarely travel abroad. In this season, highest numbers are found in lowland and coastal areas, especially on farmland, where they feed in mixed flocks with other finches. Food shortages may encourage some birds to move farther and in larger numbers.

The greenfinch's heavy bill means that it can eat larger and more varied seeds than many other finches. Weed seeds, cereals, the seeds of yew, elm and hornbeam, and rose hips and blackberries are eaten either directly from the plant or from the ground.

A common garden finch, the greenfinch has a dumpy body and stout bill. The male is green above with yellow-green underparts. Both males and females have a yellow stripe along the edge of the wing and yellow edges to the base of the forked tail. Greenfinches tend to gather in bands with other finches, in woodlands, parks and gardens. Aggressive feeders, they have learned how to cling onto peanut feeders, where they will often push the tits from pole position.

GOLDFINCH *Carduelis carduelis*

A slim, brightly coloured finch, the goldfinch has a bright red face, white cheeks and a black crown and collar. The black wings have a broad yellow stripe, very noticeable in flight, as is the white rump. Goldfinches have fine bills enabling them to feed on small seeds; winter flocks can often be seen on thistle-heads, extracting the seeds. They have a high, tinkling song, perched high in trees or in flight.

FACTS AND FEATURES

widoowit-widoowidoowit-widoowit...

Song A tinkling, twittering series of phrases, *widoowit-widoowidoowit.*
Behaviour Sings from a tree or in flight.
Habitat Gardens, orchards, open woodland, fields and waste land.
Nest Of moss, roots, lichens and thistledown high on a branch. Five or six eggs, two broods.
Food Thistle and other weed seeds, birch and alder catkins. Some beetles, caterpillars and aphids.

THE GOLDFINCH IS OFTEN regarded as the most attractive of European finches, with its smart red face and striking black-and-yellow wings. A dainty and often rather shy bird, it can be seen feeding in flocks. It has an undulating flight during which its bright yellow wing bands are the most noticeable feature. When feeding it may be restless, fluttering from one seed head to the next and calling most of the time. This agile finch can pick out small seeds from a hanging position. A group of goldfinches is known as a 'charm', which alludes to their twittering calls as they flock together. They are found along woodland edges, farmland, parks, orchards and gardens, often feeding in waste ground and on field edges where there are plenty of weeds.

The singing prowess of the goldfinch meant it was kept widely as a cage bird, with more than 100,000 caught annually in Sussex at the turn of the century.

The song is a liquid twittering, each phrase made up of tinkling notes: *widoowit-widoowidoowit-widoowit.* Song is given both from a tree, often the topmost branches, and when in flight; it is most frequently heard from March to July. The commonest call has notes similar to the song, *switt-witt-witt, swit-oolit,* uttered in flight and when perched. Another call is a grating *geez.*

The nest is built on a high fork of a tree, often a fruit tree or chestnut. Several pairs may breed together in a loose colony. Five or six eggs are laid, starting later than most finches to synchronize the hatching of the young with a good supply of suitable seeds. Two broods are usually reared.

Goldfinches are resident and migratory. Most British birds migrate to the Continent in September and October, returning in spring. Northern European goldfinches migrate south to the Mediterranean, where large concentrations of birds can occur. Late last century, goldfinch numbers in Britain declined dramatically due to trapping during their autumn migration. Then the newly formed Society for the Protection of Birds successfully campaigned for laws to protect them, and they have increased in numbers and range during this century.

The thin bill of a goldfinch enables it to get at small, compact seed heads such as those of thistle and teasel. It also feeds on other weed seeds, birch and alder catkins and some beetles, caterpillars and aphids. This bird does not visit bird tables, but if a small patch of thistles is allowed to grow at the bottom of a garden, this is likely to attract goldfinches.

SISKIN *Carduelis spinus*

IN RECENT YEARS this finch has become increasingly common in gardens during the winter. A small bird with a yellow rump, it is sometimes confused with the serin. The male has a black forehead and chin and dark wings with yellow wingbars. In the breeding season it inhabits coniferous and mixed woodland and is widespread in Scotland, in native Scots pine. It also breeds in newly afforested areas and sometimes in exotic conifers. In winter the siskin inhabits mixed woodland containing birch, larch, spruce and alder. It is a gregarious finch, feeding in small parties and sometimes associating with redpolls.

The song is a twittering warble with wheezing notes and repeated phrases, *chichi-zoooee, scurradee, scurradee, scurradee*. It is given from a perch in a tree. Early in the breeding season it can also be heard during a song flight above the trees, the male flying in circles with feathers fluffed out and exaggerated wingbeats. Singing is mainly from late February to the end of April. Flocks of siskins habitually give a flight call, a clear *tzyeee* or *tsweee*.

The nest is built near the end of a high conifer branch. From three to five eggs are laid and a second brood may be raised when there is sufficient food.

Siskins breed mainly in central and northern Europe, visiting the south and west in winter. In Britain they are found mostly in Scotland but have colonized new conifer plantations in England, Wales and Ireland. In winter, breeding birds move to the lowlands of Scotland, while some continental birds arrive in England and Wales. During hard winters many of the Scottish birds travel to England, and when food supplies on the Continent run out large numbers of birds there do likewise. They feed chiefly on alder, birch and larch cones, hanging from small branches to extract the seeds. They also forage on the ground, picking up fallen seeds. The recent trend for siskins to feed in gardens is possibly because they are attracted by seeds from ornamental conifers such as cypresses. They are also increasingly common at bird tables, eating fat and peanuts. They seem to prefer peanuts suspended in mesh bags, to which they can easily cling as their slim bills pick out the nut fragments.

FACTS AND FEATURES

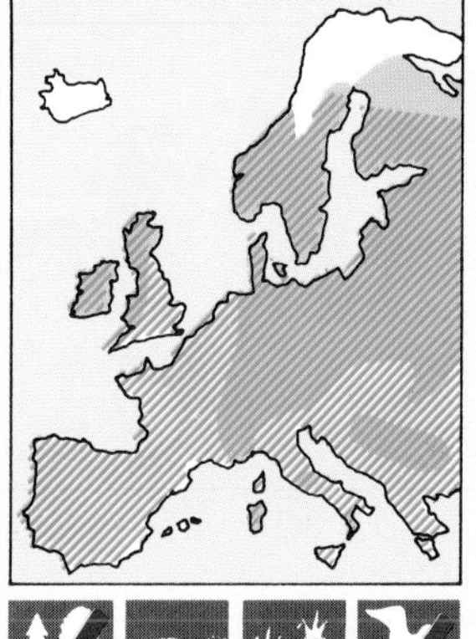

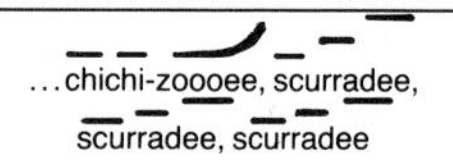

Song A lively twittering warble with repeated phrases, often ending in a wheezing note.
Behaviour Sings from a high tree perch or in song flight.
Habitat Coniferous woodland, especially spruce; birch and alder woods in winter.
Nest Of twigs, lichens and moss near the end of a high conifer branch. Three to five eggs, two broods.
Food Spruce and pine seeds, birch and alder catkins, thistle and other weed seeds. Takes fat and peanuts from feeders.

A recent visitor to gardens, this small finch is the latest to take advantage of bird tables. Like the larger greenfinch, it has quickly discovered how to get at peanuts and is even more adept at it, being used to hanging from branches to get at alder seeds. The male is yellow-green with a black cap and chin. Both sexes have two yellowish wing bars and yellow rumps.

LINNET *Acanthis cannabina*

THE MALE LINNET is a very smart-looking bird in spring, with his grey head, chestnut back, and pinky-red forehead and breast. In winter plumage he can be less easy to identify, but he has a silver edge to the closed wing and tail all year round. Linnets inhabit woodland edges where there are bushes and scrub, also downs and heaths, farmland with hedgerows, conifer plantations and gardens. On moors and in coastal areas they breed in ground vegetation such as heather and sea purslane. In winter these birds are found in open areas such as agricultural land and coastal saltmarshes.

The linnet, like the goldfinch, was a favourite cage bird and its population was reduced in the last century by widescale trapping. It is a lively finch, often seen perched on the top of a clump of gorse, calling and singing, or feeding on the ground in a flock that could number more than 200. These flocks always fly in tight packs, sometimes breaking into two and circling before they merge and land.

The song is a loud and lively musical twittering with many trills and fluty notes – *wee-cha-sii-cha-witcha-witcha-seeeee*. It is usually given from a perch at the top of a gorse or other bush, often in an excited chorus of singing birds. The main song period is from March to July, but it is often heard earlier. The flight call is a rapid, metallic *chichichichit* and the other main call is an anxious *tsooeet*.

Nests are built communally or singly in thorn bushes, small conifers, hedges and shrubs. An ornamental evergreen in a rural garden may be chosen. Between four and six eggs are laid from mid-April and early breeding allows some pairs to raise three broods.

Linnets are both resident and migratory. They are summer visitors to Scandinavia and eastern Europe and some of these birds come to Britain in winter. British linnets migrate to France and Spain in the autumn, returning in April, The diet consists almost entirely of weed seeds with a few caterpillars and other insects. Seeds of brassicas like charlock, as well as chickweed, persicaria and fat-hen, are the most popular foods in England.

In many upland areas the linnet is replaced by its close relative the twite, *Acanthis flavirostris*. The twite's song is more metallic and its call is a distinctive nasal *chweek* which can be given in flight.

FACTS AND FEATURES

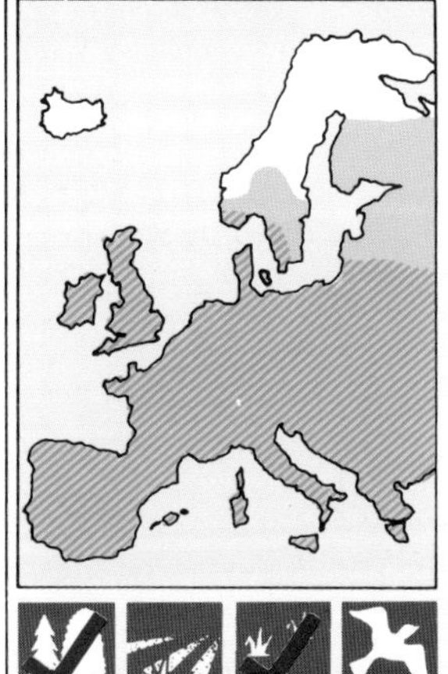

wee-cha-sii-cha-witcha-witcha-seeeee

Song A musical twittering with fluty and trilling notes.
Behaviour Sings from the branch of a tree or bush and in flight.
Habitat Commons and rough ground with bushes and scrub, plantations and gardens. Fields and coasts in winter.
Nest Of grass, moss, twigs and roots in evergreens, gorse and other shrubs. Four to six eggs, two or three broods.
Food Weed seeds, particularly charlock, some caterpillars and other insects.

A common countryside bird which can be seen in gardens, the linnet is more often seen in winter flocks, feeding in open agricultural land. The male has a grey head, chestnut back and a red breast and forehead. The female is duller-coloured, and streaked; both sexes show silvery-white edges to the wing and tail. The linnet has a loud and lively tuneful song.

REDPOLL *Acanthis flammea*

This is a small streaky-brown finch with two pale buff wing bars. In spring, males have a red forehead and pink breast and both sexes have a small yellowish bill and a black chin. They feed in flocks, hanging from the branches of birches and aspens, and have a loud buzzing flight call. Flocks can be extremely noisy as they flit from one tree to another.

FACTS AND FEATURES

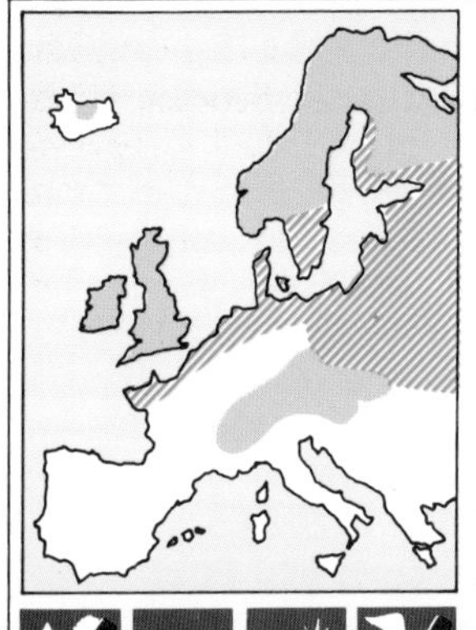

trrrrr, chee-chee-chee

Song A trill, mixed with its chattering flight call, *trrrrr, chee-chee-chee.*
Behaviour Sings during song flight and also when perched.
Habitat Birch and alder copses, conifer plantations, parks and gardens.
Nest Of twigs and stalks in bushes and young conifers. Four or five eggs, one or two broods.
Food Birch, alder, sallow and weed seeds with some insects. Occasionally takes peanuts from feeders.

THIS SMALL, BROWN FINCH is found in mixed woodland of birch, alder, willow and conifers, and in parkland and gardens. The planting of many areas with conifers has encouraged its spread in Britain. It is often seen flying at some height, giving its trilling call. In winter it forms into flocks, sometimes numbering dozens of birds. They can appear flighty, suddenly taking to the air in a cloud, circling and landing again.

Redpolls feed in trees and on the ground and they can be very approachable. In winter their streaky brown plumage appears rather dull, but in spring the male acquires a brilliant red forehead and bright pink breast.

The song is a mixture of twitters, trills and call notes: *trrrrr, chee-chee-chee, trrrrr, chee-chee-chee.* The bird sings from a perch and also in song flight, circling the tree tops in a bouncy manner. The main song period is from mid-March to mid-August. The flight call is a staccato, buzzing *chee-chee-chee, chee-chee* and there is also a plaintive *tsooeet.*

Nests are often built communally in small trees and tall bushes, particularly birch, alder, sallow, hawthorn, gorse and young conifers. Four or five eggs are laid and one or two broods reared.

Redpolls are northern finches, known only as winter vistiors in much of Europe; they are the only finches to breed in Iceland. They are a variable species and occur in a number of races. Northern Scandinavian birds, which are summer visitors to central and parts of western Europe, are larger and paler and are called mealy redpolls. They have white rather than buff feathers with whiter wing bars and a pale rump, and occur irregularly in Britain, often on the east coast.

Redpolls feed mainly on seeds: in spring sallow seeds and some insects, in summer grass and weed seeds, and in autumn and winter birch and alder seeds. In hard winters they may visit bird tables, sometimes hanging at peanut feeders in the manner of siskins.

CROSSBILL *Loxia curvirostra*

THE CROSSBILL IS one of the largest finches, stout and with a large head, long wings and short tail. Its bill is adapted for extracting seeds from conifer cones, being crossed over at its tip. The male is crimson and orange on the head, underparts and rump, with dark brown wings and tail. Crossbills are rarely seen since they spend most of their time hidden in conifers, and the only sign may be the sound of cones being cracked to extract seeds. These birds breed in conifers and have benefited from new plantations. They are usually seen on the ground only when drinking, but may descend for seeds from fallen cones. Crossbills are most likely to be noticed in spring, when noisy family groups fly from tree to tree in search of food.

The song is not very complex, being made up of a series of chipping calls that almost form a trill, followed by a two-syllable note: *chip-chip-chip-chip-chip-ticheee*. Redstart-like warblings may also be heard in between song phrases. Crossbills use song posts infrequently between January and July. The call note is an explosive *chip*, repeated constantly when the bird is in flight and also when it is perched.

Nests are built high in conifers, usually at the edge rather than in the middle of a clump. These birds can breed as early as January and February, and breeding has been recorded in every month of the year. Four eggs are laid and where food is plentiful a second brood may be reared.

Crossbills are found all over Europe, particularly in higher southern regions. Northern birds are subject to 'irruptions' when their food supply becomes scarce, large numbers arriving in Britain during autumn. British crossbills are scattered across Scotland, England and Wales in the breeding season. In winter they remain near their breeding areas, with continental birds adding to their numbers; they also appear in areas where there are no breeding birds. They feed mainly on spruce but also on pine, larch and birch. When conifer seeds are exhausted they have been known to eat oak buds, thistle seeds, berries and insects such as aphids.

A small isolated population in the Scottish Highlands is sufficiently distinct to be called a separate species. This is the Scottish crossbill, *Loxia scotica*, and it is the only species of bird found exclusively in Britain. It has a thicker, heavier bill than the crossbill and a stronger *joop, joop* call.

FACTS AND FEATURES

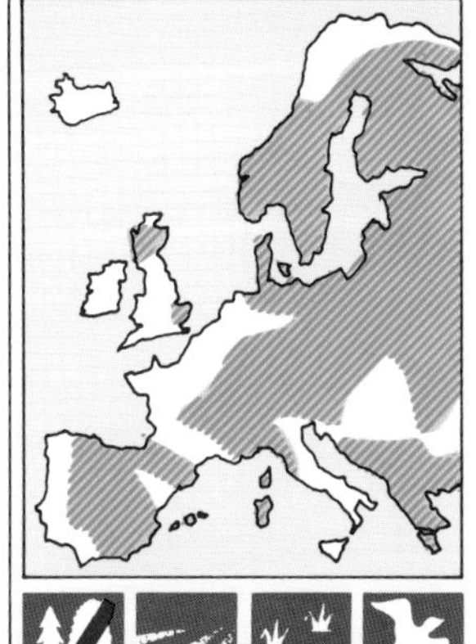

chip-chip-chip-chip-chip-tichee...

Song A succession of call notes followed by a loud *ticheee* note.
Behaviour Sings from an exposed branch or the top of a tree.
Habitat Coniferous forests of pine, larch and spruce.
Nest Of pine twigs, grass and wool in a clump or belt of conifers. Four eggs, one or two broods.
Food Seeds of pine, larch and spruce; oak buds and thistle seeds; ivy, hawthorn and rowan berries; some aphids, beetles and caterpillars.

This is the largest finch, with long wings, a short tail and a large head. Males are brick red with dark brown wings and tail; females are olive green. To enable it to extract seeds from conifers it has developed a peculiar bill in which the mandibles cross over. When it is feeding high in a pine tree, a cracking noise can be heard, accompanied by falling fragments of pine cones.

BULLFINCH *Pyrrhula pyrrhula*

THE MALE BULLFINCH is a striking bird with his pinky-red underparts and cheeks, and black cap and wings. In flight the most distinctive feature is a broad white rump that contrasts with the grey back and black tail. Bullfinches are dumpy birds with short, deep bills and are usually found in pairs or family groups along woodland edges, orchards, young plantations, parks and gardens. In some parts of Britain they damage apple and pear trees by taking the buds, being regarded as pests.

Bullfinches were once common in captivity and trained to sing. With no complex song of their own they could easily learn tunes such as 'God Save the King'! The wild song is a soft, creaky warble, *tee-tee-teeoo*, seldom heard and given by both males and females from March to June. The familiar call is a sad, piping *deeu*, often used as a contact note when birds have become separated. It is a loud, currying call that can be uttered when perched or in flight.

A beautiful but sometimes destructive bird, the bullfinch has a great fondness for fruit buds in spring. The male has red underparts, the female pinky-brown ones. Both have a black crown, grey back, black wings and white wing bar, and a black tail with a bold white rump which is distinctive in flight. Young birds resemble females, but lack the black crown.

FACTS AND FEATURES

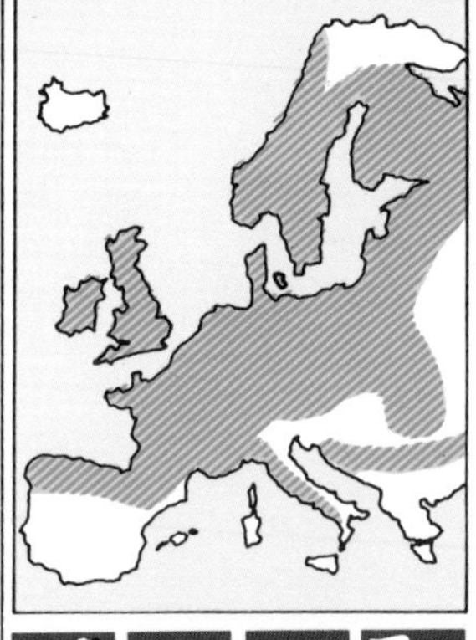

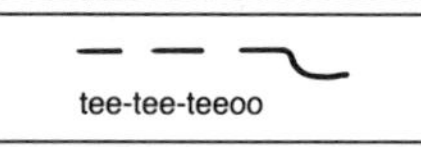

Song A soft, creaky warble, *tee-tee-teeoo*.
Behaviour Sings quietly from bushes and trees.
Habitat Woodland, thickets, hedgerows, parks, gardens and orchards.
Nest Of thin twigs, moss and lichens in a thick hedge or bush. Four or five eggs, two broods.
Food Buds of apple, pear and hawthorn, seeds of ash, elm and dock, berries from bramble and mountain ash.

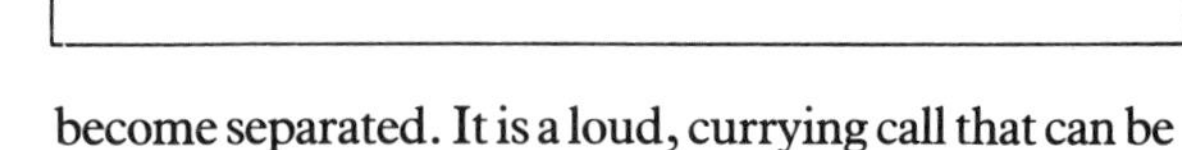

Nests are always well hidden in the depth of a thick hedge, bush or evergreen. Four or five eggs are laid and two broods are raised.

Bullfinches are resident across most of Europe, except for parts of the south and east, where they may occur in winter, and in northern Scandinavia, where they are absent or summer visitors. Scandinavian birds are a distinct race, larger and brighter with a paler back; they are scarce winter visitors to Britain, mainly on the north and east coasts of Scotland and England. They feed on buds and flowers from fruit trees, hawthorn, elm, blackthorn, sallow and oak, and also on seeds from ash, elm, birch, bramble, dock and many other weeds. Caterpillars and spiders may be eaten or fed to the young.

A large European finch that shares similarities with the bullfinch is the pine grosbeak, *Pinicola enucleator*; it is confined to Scandinavia. It is pinky-red with a grey back and two pale wing bars, and the call is a bullfinch-like *tee-tee-tew*.

The scarlet rosefinch, *Carpodacus erythrinus*, is a small, red-coloured finch that also breeds in parts of Scandinavia and in eastern Europe. It has been expanding its range westwards and has recently bred in Britain. The song of the scarlet rosefinch consists of a piping of four to six notes, and this finch's call is a piping *tew-it*.

HAWFINCH *Coccothraustes coccothraustes*

THE HAWFINCH is the least seen of all British finches. Secretive and wary, it spends most of its time in the tops of woodland trees. It is a sizeable bird with a massive bill, large head, thick neck, stout body and short tail, making it distinctive when both perched and flying. It has a fast undulating flight, when its pale wing patches can be seen. Habitually the hawfinch perches at the tops of trees, although it comes to the ground to feed on fallen seeds, hopping or walking with a waddle. These birds live in mature deciduous woodland with trees like beech, hornbeam, wych elm, sycamore and maple. Orchards, parks and gardens are also frequented, especially if there are cherry trees. The hawfinch can be easier to see in the winter, when it descends more to the ground to feed, and in hard weather it visits gardens more frequently.

The song, which is rarely heard, is a high and halting warble, *teek, teek, tur-whee-whee*. It is usually delivered from the top of a high tree between March and May. More noticeable is the loud *tic* or *tsik* call note, often given twice in quick succession and reminiscent of a robin. The bird calls when perched and in flight.

Nests are usually built on horizontal tree branches high off the ground, but they are sometimes sited lower in saplings. Pairs may nest close together and birds usually feed away from their breeding territories. Four or five eggs are laid and one brood is usually raised, but two or three if breeding starts early.

Hawfinches breed from Europe eastwards to Japan, but they are absent from northern Europe and Asia. In the north of their range they are migratory and move south to winter in southern Europe. Some of these migrants reach Britain, but many are overlooked.

These finches feed on large seeds and fruit stones that their huge, powerful bills crack open with ease. In woodland they like seeds of wych elm, beech, hornbeam and sycamore. Hedges are raided for berries such as hips, haws and holly. Cherry orchards are a favourite haunt and the noise of cracking stones is distinctly audible. Pips are also extracted from fallen apples, and some insects such as beetles and caterpillars are taken, especially when feeding young. During hard weather hawfinches can be seen in parks and gardens, eating the more exotic fruits of cotoneaster, berberis and honeysuckle.

FACTS AND FEATURES

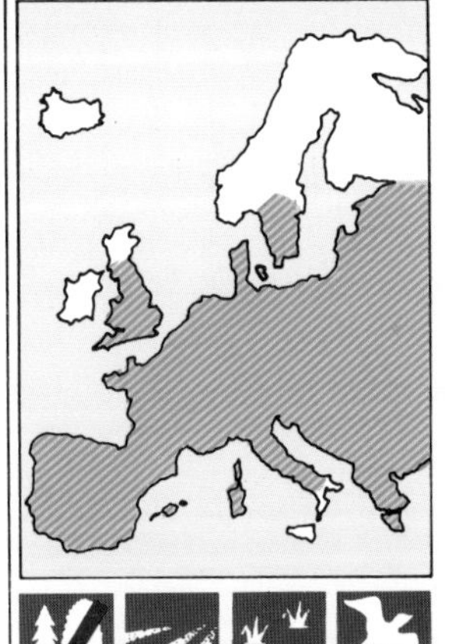

teek, teek, tur-whee-whee

Song An infrequently heard, halting *teek, teek, tur-whee-whee*.
Behaviour Sings from the top of a tree.
Habitat Mature deciduous and mixed woods, gardens, parks and orchards.
Nest Of twigs, roots and lichens on a horizontal tree branch. Four or five eggs, one or occasionally two or three broods.
Food Kernels and seeds such as hornbeam, elm, cherry, apple, beech, hawthorn and yew. Also some caterpillars and beetles.

This large, stocky finch is difficult to see for most of the year, keeping high in woodland trees. In winter it is more likely to be found feeding on the ground, where its large head, stout body and huge bill are unmistakable. Its bill is powerful enough to split cherry stones with ease. In flight it shows white wing patches and a white tip to its short tail.

YELLOWHAMMER *Emberiza citrinella*

THE MALE YELLOWHAMMER is a distinctive bird as he sits on the top of a hedge or bush, with his bright yellow head and underparts. Even as he flies away he is readily recognizable by his bright, rust-coloured rump and white outer tail feathers. Like most buntings, the yellowhammer is a fairly slim, long-tailed bird which spends much time feeding on the ground. It is sociable and flocks with others of its kind, as well as with sparrows and finches, to feed on stubble fields and other agricultural land. Flocks of tens or even hundreds roost together in dense, scrubby vegetation. The species is found along woodland edges, in young conifer plantations and hedgerows and in heaths, commons and scrub-covered hillsides. It prefers open country with bushes and hedges that can be used as song posts.

The song is a familiar sound of the British countryside and the rhythm of the series of high notes is commonly written as 'a-little-bit-of-bread-and-no-cheese'. In reality there is usually one less syllable: *chi-chi-chi-chi-chi-chi-chweeeee*. The song is given from a prominent perch on a bush, small tree, hedge or telephone wire. Males begin to sing in February and continue until the end of August. Calls, when perched and in flight, include a ringing *twink* and a sharp *twik* or *twitik*.

Nests are built low in bushes, hedges and young trees. The first clutch of three or four eggs may be laid in late April, leaving time for two or three broods.

Yellowhammers are resident over most of Europe, but northern birds migrate south and parts of the

Perched on top of a bush the bright yellow male is hard to overlook. Its famous 'little-bit-of-bread-and-no-cheese...' song can be heard from song posts in hedgerows and at woodland edges. In flight the yellowhammer shows a bright rusty-coloured rump and white outer tail feathers. Females are duller and more streaked.

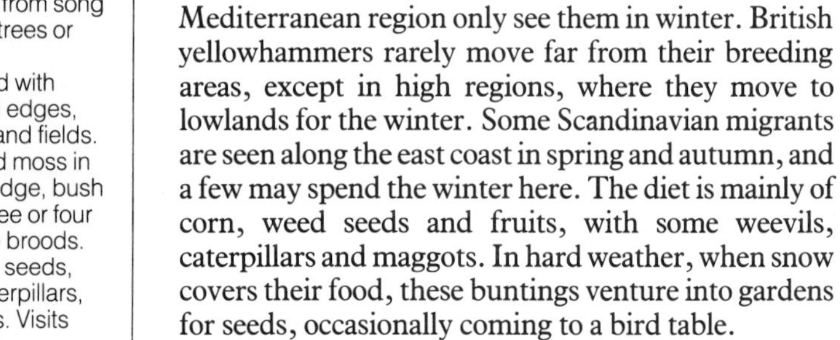

Mediterranean region only see them in winter. British yellowhammers rarely move far from their breeding areas, except in high regions, where they move to lowlands for the winter. Some Scandinavian migrants are seen along the east coast in spring and autumn, and a few may spend the winter here. The diet is mainly of corn, weed seeds and fruits, with some weevils, caterpillars and maggots. In hard weather, when snow covers their food, these buntings venture into gardens for seeds, occasionally coming to a bird table.

FACTS AND FEATURES

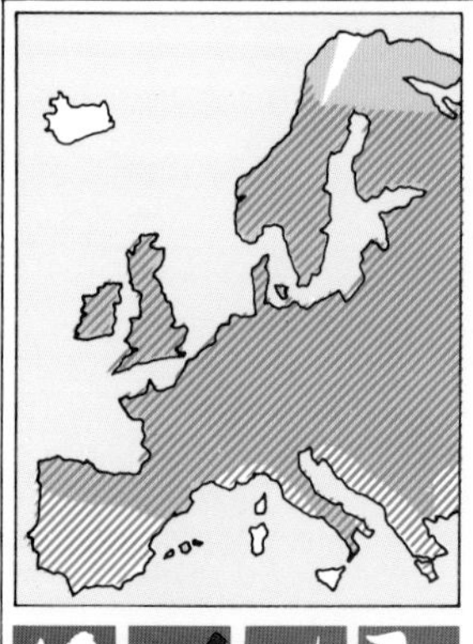

chi-chi-chi-chi-chi-chi, chweeeee

Song A repeated high note followed by a longer one.
Behaviour Sings from song posts on bushes, trees or wires.
Habitat Grassland with hedgerows, wood edges, bushy commons and fields.
Nest Of stalks and moss in the bottom of a hedge, bush or young tree. Three or four eggs, two or three broods.
Food Corn, weed seeds, fruits, weevils, caterpillars, grubs and spiders. Visits gardens for seeds.

CIRL BUNTING *Emberiza cirlus*

A southern European relative of the yellowhammer with yellow cheeks, a black throat and dark crown and eye-stripe. In flight it can be distinguished from the yellowhammer by its olive-coloured rump. In Britain it is at the northern limit of its range and has become confined to a few southern counties. It has a monotonous jangling song rather like the lesser whitethroat's.

SIMILAR IN SIZE to the yellowhammer, the male cirl bunting has a black throat and eye-stripe with a greyish crown and neck. In flight it is difficult to distinguish from the yellowhammer unless the olive-coloured rump is seen. These birds are found in open country with hedges and trees. In southern Europe they inhabit valleys and hillsides with scattered bushes, vineyards and orchards. They can be difficult to see, often sitting on a low bush or in scrub and not moving until they fly off. Cirl buntings spend a lot of time feeding on the ground, sometimes in the company of yellowhammers and finches.

The song is a rather monotonous jangling trill on a single note, rather similar to a lesser whitethroat's song: *ji-i-i-i-i-i-i*. An exposed elevated song post is usually chosen, such as a branch of a tree, the top of a bush or a telephone wire. Song can be heard all year round but mainly from late February to early September. The call note is a thin *sip* or *sii*.

The nest is situated low in a dense hedge or thicket of bramble, sometimes on the ground. Three or four eggs are laid and two broods reared. The young stay with their parents after fledging and family parties can be seen feeding together in autumn.

Cirl buntings are a southern European species which just reaches north into southern England. Their numbers here have declined dramatically in recent years, perhaps due to climatic factors, and about 50 pairs were known in 1985. A similar decline has also occurred in Belgium and northern France. They are residents throughout Europe, but wander from their breeding areas in winter. They feed mainly on corn, weed and grass seeds, some beetles and caterpillars, as well as blackberries, haws and elderberries.

FACTS AND FEATURES

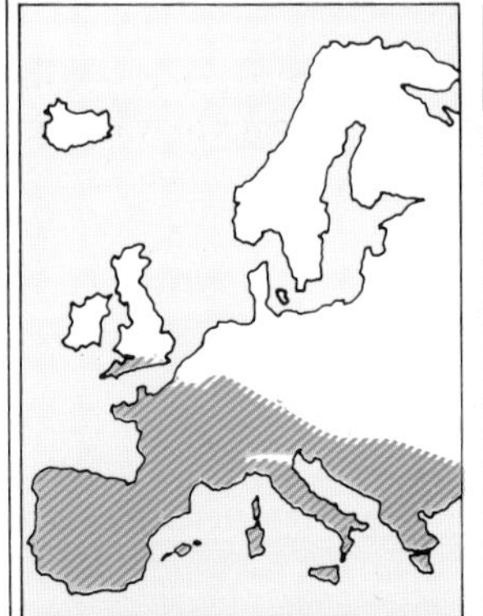

ji-i-i-i-i-i-i-i-i-i...

Song A jangling repetition of a single note, rather like a lesser whitethroat.
Behaviour Sings from the top of a hedge, bush, wall or wire.
Habitat Pasture and downland with bushes and trees, bushy hillsides.
Nest Of roots, moss and leaves low in a thick hedge or bush. Three or four eggs, two broods.
Food Corn, weed and grass seeds, beetles and caterpillars, some berries.

ROCK BUNTING *Emberiza cia*

The male rock bunting has a chestnut belly and a grey upper breast and head, with a very distinctive black face pattern. In flight it has a rusty rump like a yellowhammer, but shows less white in the tail. Like many other buntings, the females are brown and streaked. The rock bunting can be difficult to see as it perches on a bush or rock, only its high call-note giving it away.

FACTS AND FEATURES

zitt-zitterit-zitt-zitt-zitterit-zitt

Song A rapid dunnock-like song beginning with a high *zitt*.
Behaviour Sings from an exposed branch or wire.
Habitat Bush-covered stony slopes in hilly areas.
Nest Of grass and moss on the ground among stones and dry vegetation. Four to six eggs, usually one brood.
Food Oats, and other cereals, grass seeds, caterpillars and other insects.

A DISTINCTIVELY PATTERNED bunting which can be difficult to see, this bird perches motionless on a low bush or rock; if disturbed it often flies some distance before landing. In flight it displays a chestnut rump and white outer tail feathers reminiscent of a yellowhammer. The rest of the plumage is quite unlike any other European bunting and the male has a chestnut belly, grey head and upper breast, and black-striped face. This species inhabits warm hillsides with scrub, bushes and small conifers, open rocky areas and quarries, vineyards or vegetable gardens with stone walls, river valleys and alpine meadows. In winter it is found in fields with bushes and hedgerows, frequently flocking with other buntings. The rock bunting feeds on the ground and spends the rest of its time perched in a bush, calling occasionally. It can easily creep through a bush to move position, instead of flying.

The song is a high dunnock-like warbling which commences with a *zitt* and can be written *zitt-zitterit-zitt-zitt-zitterit-zitt*. The bird sings from a high perch at the top of a tree or bush and also from posts and telephone wires. A song flight may accompany the singing, in which the male flutters into the air while spreading his tail to expose the white edging. The call is a thin, high *tsi* or *tzit* given discreetly from the top of a bush or rock. In flight, and when in a flock, a double *tzi-tzi* is sometimes uttered.

The nest is built on the ground among stones and dry vegetation. The parents land a little distance from the nest and climb over the rocks to reach it, but when leaving they fly directly away. Between four and six eggs are laid and one brood is reared, sometimes two.

Rock buntings are mainly resident and are found in southern Europe, reaching as far north as the Rhine valley in Germany. Birds that breed in mountainous areas move to lower ground in winter, and the most northerly breeding birds tend to migrate south towards the Mediterranean. This bunting has been seen in Britain on a handful of occaions as an extremely rare vagrant. The food is cereals such as oats, grass seeds, and caterpillars and other insects.

ORTOLAN BUNTING *Emberiza hortulanus*

THE ORTOLAN BUNTING'S NAME is synonymous with shooting in parts of Europe, since this summer visitor has long been regarded as a gourmet food item. In spring the male is subtly handsome with chestnut-buff underparts, a greeny head and a yellow throat and moustache. He has a white edge to his tail and a pale eye-ring makes him even more distinguished. This bunting is an unobtrusive bird, spending much of its time on the ground but also perching on bushes and small trees. In common with most members of its family it can run and creep on the ground or skulk in bushes, remaining unobserved until it flies away.

Ortolan buntings like open ground with trees, bushes and rocks, and they can be found from lowlands to mountains. They feed on grains, grass and weed seeds, beetles, caterpillars, crickets, moths and snails. Vineyards, cornfields, woodland edges and glades, copses and even gardens may be used for breeding.

The song is a series of seven or eight melancholy notes with the last ones lower: *zee-zee-zee-zeu-zeu-zeu-zeu-zeu*. The male sings from a perch near the top of a tree or bush and sometimes from a rock or telephone wire. The call note usually is a soft *tlip*, or occasionally a hard *twik* or whistled *teu*.

The nest is built in a grassy hollow or tussock, well hidden from view. Between four and six eggs are laid and two broods are commonly raised.

Ortolans breed over most of Europe with the exception of the north and parts of the west. They arrive in southern Europe in mid-April and reach southern Scandinavia by mid-May. Departure south begins in August and continues until late October. The main wintering area is in north-eastern Africa. In Britain, this species occurs as a scarce migrant in spring and autumn – mostly the latter, when it is found mainly in coastal fields.

A close relative of the ortolan is Cretzschmar's bunting, *Emberiza caesia*, whose European population is confined mainly to Greece. It has an orange throat and moustache and a greyer head. The song is a series of three or four *piu* notes and the call a hard *tsyip*.

A bird of the open countryside in most of Europe (where it is shot as game), the male ortolan bunting has a grey-green head with a yellow throat and moustache. Its underparts are pinky-brown and its back streaky brown. Its bill is pink and its dark eye is surrounded by a pale eye-ring.

FACTS AND FEATURES

zee-zee-zee-zeu-zeu-zeu-zeu-zeu

Song A series of high *zee* and lower *zeu* notes.
Behaviour Sings from a bush, rock or wire.
Habitat Rocky hillsides and low ground with trees and bushes, woodland edges, gardens and fields.
Nest Of grass and roots on the ground in vegetation. Four to six eggs, two broods.
Food Grains, weed and grass seeds, caterpillars and beetles.

REED BUNTING *Emberiza schoeniclus*

THE MALE REED BUNTING is an attractive bird, his black hood and white moustache contrasting with his otherwise rather sparrow-like appearance. He usually perches at the edge of a bush or part of the way up a swaying reed, flicking his wings or fanning his tail. This species has a low and jerky flight, showing the white outer tail feathers. When feeding on the ground it can hop, creep and run with ease. At one time it was confined to wetland areas, marshes, rivers, and ditches, but in recent years it has been colonizing agricultural land, conifer plantations and downland. In winter the male's head becomes browner and these birds flock together to feed on farmland and waste ground, and sometimes in gardens. They also roost in flocks, usually choosing a marshy area.

The simple song has two or three repeated notes and a characteristic ending, *twee, twee, twee, tititick*. The singing bird always perches on a bush, reed or other tall piece of vegetation and can sing a dozen phrases each minute. The main song period is from late March to July. Call notes include a shrill *tsee-u*, rather like a yellow wagtail's call, and a metallic *chink*.

The nest is built on or close to the ground in a tussock of grass, thick bush or hedge. Four or five eggs are laid from late April onwards and two, or even three, broods may be raised.

Reed buntings are resident in most of Europe, although Scandinavian and eastern European birds migrate in winter. Some western Scandinavian birds come to Britain at this time, while others pass down the east coast on their way farther south. British reed buntings are mainly resident; however, birds breeding in high areas, which become inhospitable in winter, move to lower ground. These buntings feed mainly on grass seeds, grain, caterpillars, small snails and beetles. During cold winters they may visit gardens and take seeds from the ground or a bird table.

Another bunting with a black head is the Lapland bunting, *Calcarius lapponicus*. It breeds in northern Scandinavia and winters along the North Sea coast. It has a short, musical song and a distinctive flight call, *tik, titik, tikitik*.

FACTS AND FEATURES

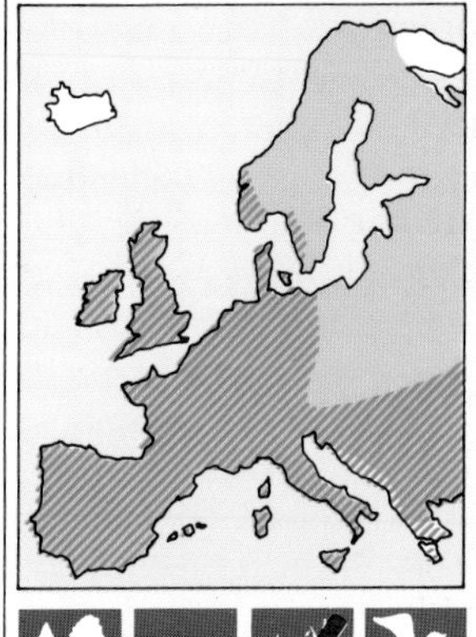

twee, twee, twee, tititick

Song A simple, deliberate and chirruping song.
Behaviour Sings from bushes, stems of reeds and other tall plants.
Habitat Reedbeds, waterside vegetation, farm hedges and conifer plantations.
Nest Of grass and moss in a bush or grassy tussock. Four or five eggs, two or three broods.
Food Grass seeds, grain and small creatures such as caterpillars. Visits bird tables for seeds in cold winters.

The male reed bunting has a distinctive black hood and white moustache which merges with its white collar and underparts. The back is streaked brown and the tail has white edges, easily seen in the jerky flight pattern. Reed buntings were once confined to wetland areas, but can now be found on agricultural land; they even visit gardens for food during severe winters. They roost in flocks, preferring marshy areas.

CORN BUNTING *Milaria calandra*

THE CORN BUNTING IS the largest of the buntings and probably the least interesting in appearance. It is a stocky bird with streaky brown plumage, the pale, streaked breast and ill-defined moustache being the only features when perched. In flight it is broad-winged and its fluttering action is made more distinctive by the habit of trailing its legs. When not feeding, the corn bunting spends most of its time perched on a bush, hedge, fence or telephone wire, from where it delivers its song. It is a bird of arable land with hedges and fences, and one of the few birds to have benefited over most of its range from agricultural changes that have removed so much woodland and other habitats. However, in recent years the British population has declined – probably due to modern arable farming techniques, which have removed crop rotation, and the widespread use of herbicides. It keeps a well-defined territory, but is communal in winter, with roosts of up to 500 birds not uncommon.

The song is a loud, jangling series of accelerating notes that ends with a sound like a bunch of keys being shaken: *tick, tick, tick, tick-tick-tick-tchweeeeeee*. Singing is from territorial song perches on posts, wires, walls and bushes. In areas where corn buntings are common, they can be seen dotted at regular intervals along telephone wires. The main song period is from mid-February to mid-August. In flight, this bird has a harsh *chip* call, and flocks may give a loud *quit*.

Nests are built in thick vegetation, brambles, bushes and corn, sometimes off the ground. From three to five eggs are laid, and since the breeding season does not begin until late May, it is not always possible to raise two broods.

Corn buntings are resident over most of their range, moving only short distances in winter. In eastern Europe they are more migratory and birds from there spend the winter in Egypt. In Britain, these buntings are absent from upland areas and are mainly concentrated along the eastern side, although they are scarce in East Anglia. In winter there is some evidence that they move nearer to the coast. These buntings eat weed and grass seeds, grains, grasshoppers, beetles, caterpillars, spiders and worms. In hard winters, birds may venture into rural gardens to take seeds from the ground, often in the company of sparrows or finches.

FACTS AND FEATURES

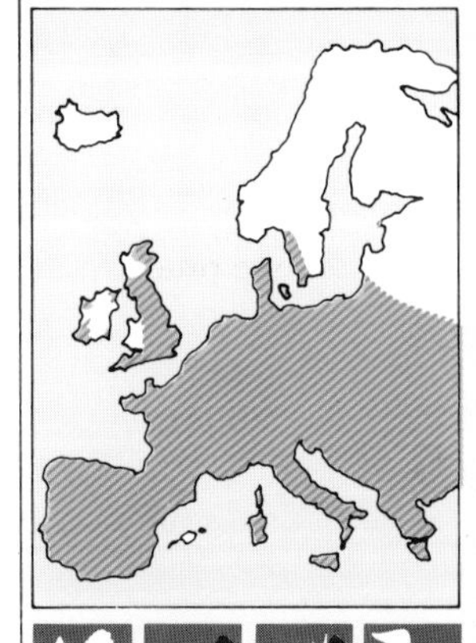

tick, tick, tick, tick-tick-tick-
tchweeeeeee

Song A series of accelerating ticking notes with a jangling finish.
Behaviour Sings from a song post on a bush, hedge or wire.
Habitat Arable and pasture land with hedges and bushes.
Nest Of grass and stems in tall vegetation, brambles and hedges. Three to five eggs, one or two broods.
Food Weed and grass seeds, corn and other grains, grasshoppers, beetles, caterpillars, spiders and worms.

A large brown bunting with streaky brown plumage, mostly seen perched on a bush, hedge or telephone wire. In flight it shows no white in the tail or wings and has the habit of trailing its legs. Both sexes are identical. Its loud jangling song is a common sound in open grassland and some agricultural areas.

USEFUL ADDRESSES

ENGLAND

Royal Society for the Protection of Birds
The Lodge
Sandy
Bedfordshire
SG19 2DL

British Trust for Ornithology
Beech Grove
Tring
Hertfordshire
HP23 5NR

International Council for Bird Preservation
32 Cambridge Road
Girton
Cambridge
CB3 0PJ

SCOTLAND

Scottish Ornithologists' Club
21 Regent Terrace
Edinburgh
EH7 5BT

EIRE

Irish Wildbird Conservancy
Ruttledge House
8 Longfield Place
Monkstown
Co Dublin

BELGIUM

Comité de Coord pour la Protection des Oiseaux
Durentijdlei 14,
2130 Braaschaat,
Antwerp

DENMARK

Dansk Ornithologisk Forening
Vesterbrogade 140
1620 Copenhagen

FRANCE

Ligue Francaise pour la Protection des Oiseaux
La Corderie Royale
BP 263
F 17315 Rochefort

GERMANY

Deutscher Bund fur Vogelschütz
Achalmstrasse 33A
D 7014 Kornwestheim

GREECE

Hellenic Ornithological Society
PO Box 64052
GR-15701
Zographos

ITALY

Lega Italiana Protezione Ucelli
Vicolo San Tiburzio SA
43100 Parma

NETHERLANDS

Nederlandse Vereniging tot Bescherming van Vogels
Dreibergseweg 16c
3708 JB Zeist

NORWAY

Norsk Ornithologisk Forening
Innherredsveien 67A
7000 Trondheim

PORTUGAL

Ligu Para a Proteccao da Natureza
Estrada do Calhariz de Benfica 187
1500 Lisbon

SPAIN

Sociedad Espanola de Ornitologia
Facultad de Biologia
Pl 9
28040 Madrid

SWEDEN

Sveriges Ornitologiska Forening
Runebergsgaten 8
11429 Stockholm

AMERICA

American Birding Association
PO Box 4335
Austin
TX 78765

American Ornithologists Union
National Museum of Natural History
Smithsonian Institution
Washington
DC20560

AUSTRALIA

Royal Australian Ornithologists Union
21 Gladstone Street
Moonee Ponds
Victoria 3039

CANADA

Canadian Nature Federation
Suite 203
75 Albert Street
Ottawa
K1P 6G1

INDIA

Bombay Natural History Society
Hornbill House
Shadid Bhagat Singh Road
Bombay 400 023

JAPAN

Wild Bird Society of Japan
Aoyama Building
1-1-4 Shibuya
Shibuya-ka
Tokyo 150

Index

Credits

Photographs by kind permission of the Frank Lane Agency.